Studies in Fuzziness and Soft Computing

Volume 352

Series editor

Janusz Kacprzyk, Polish Academy of Sciences, Warsaw, Poland
e-mail: kacprzyk@ibspan.waw.pl

About this Series

The series "Studies in Fuzziness and Soft Computing" contains publications on various topics in the area of soft computing, which include fuzzy sets, rough sets, neural networks, evolutionary computation, probabilistic and evidential reasoning, multi-valued logic, and related fields. The publications within "Studies in Fuzziness and Soft Computing" are primarily monographs and edited volumes. They cover significant recent developments in the field, both of a foundational and applicable character. An important feature of the series is its short publication time and world-wide distribution. This permits a rapid and broad dissemination of research results.

More information about this series at http://www.springer.com/series/2941

Arindam Chaudhuri · Krupa Mandaviya
Pratixa Badelia · Soumya K. Ghosh

Optical Character Recognition Systems for Different Languages with Soft Computing

Springer

Arindam Chaudhuri
Samsung R&D Institute Delhi
Noida, Uttar Pradesh
India

and

BIT Mesra Patna Campus
Patna
Bihar
India

Krupa Mandaviya
MEFGI Engineering College
Rajkot
Gujrat
India

Pratixa Badelia
MEFGI Engineering College
Rajkot
Gujrat
India

Soumya K. Ghosh
Department of Computer Science
  Engineering
School of Information Technology
Indian Institute of Technology
Kharagpur
West Bengal
India

ISSN 1434-9922   ISSN 1860-0808   (electronic)
Studies in Fuzziness and Soft Computing
ISBN 978-3-319-84357-5   ISBN 978-3-319-50252-6   (eBook)
DOI 10.1007/978-3-319-50252-6

Printed on acid-free paper

This Springer imprint is published by Springer Nature
The registered company is Springer International Publishing AG
The registered company address is: Gewerbestrasse 11, 6330 Cham, Switzerland

To our families and teachers

Contents

List of Figures

List of Tables

Chapter 1
Introduction

1.1 Organization of the Monograph

Optical character recognition (OCR) is one of the most popular areas of research in pattern recognition [3, 25] since past few decades. It is an actively studied topic in industry and academia [8, 15, 18, 24] because of its immense application potential. OCR was initially studied in early 1930s [23]. It has its origins in Germany as a patent by Gustav Tauschek. OCR is a technique of translating handwritten, typewritten or printed text characters to a machine-encoded text [23, 25]. It is a process of reading handwritten characters and recognizing them. It is widely used as a form of data entry from printed paper data records which may include passport documents, invoices, bank statements, computerized receipts, business cards, mail, printouts etc. OCR is also recognized as the subdomain of image processing which is an important research area of pattern recognition. The human brain generally finds some sort of relation predominantly in graphical form in order to remember it and recognize later. In a way it tends to produce or find patterns in handwritten characters. This led to the major motivation towards the development of OCR systems. The characters of various available languages are based on the lines and curves. An OCR can be easily designed to recognize them.

The image captured by digital camera is converted into a suitable form required by the machine. It is a common method of digitizing printed texts so that it can be electronically edited, searched, stored more compactly, displayed on-line and used in machine processes such as machine translation, text-to-speech, key data and text mining. Availability of huge datasets in several languages has created an opportunity to analyse OCR systems analytically. The major concern in these systems is the recognition accuracy. There is always an inherent degree of vagueness and impreciseness present in reallife data. Due to this recognition systems are treated here through fuzzy and rough sets encompassing indeterminate uncertainty. These uncertainty techniques form the basic mathematical foundation for different soft computing tools. A comprehensive assessment of the methods are performed through English, French, German, Latin, Hindi and Gujrati languages [8]. The experiments are performed on several publicly available datasets. The simulation

© Springer International Publishing AG 2017
A. Chaudhuri et al., *Optical Character Recognition Systems for Different Languages with Soft Computing*, Studies in Fuzziness and Soft Computing 352, DOI 10.1007/978-3-319-50252-6_1

studies reveal that the different soft computing based modeling of OCR systems performs consistently better than traditional models. The results are evaluated through different performance metrics. Several case studies are also presented on the abovementioned languages to show the benefits of soft computing models for OCR systems.

This book is written in the following divisions: (1) the introductory chapters consisting of Chaps. 1 and 2 (2) soft computing for OCR systems in Chap. 3 (3) OCR systems for English, French, German, Latin, Hindi and Gujrati languages in Chaps. 4–9 and (4) summary and future research in Chap. 10. First we become familiar with OCR. All we need to know about OCR for this book comprises of Chap. 2. A beginners' introduction to OCR is given in [23]. This is followed by a discussion on the different soft computing tools for OCR systems in Chap. 3 [8]. The soft computing tools highlighted in this Chapter is based on the concepts given in book [12] where reader can refer the introductory soft computing concepts. This Chapter forms the central theme of the entire book. A better understanding of this Chapter will help the reader to apply the stated concepts in several OCR applications encompassing different languages. This Chapter is very important for conceptually appreciating the soft computing methods and techniques for different languages presented in Chaps. 4–9.

The six different languages considered in this monograph are the most widely spoken language across the world. The character set used in these languages have a great degree of variation among them. Thus it would be quite unfair to place all the languages as subsections in one Chapter. This would make the situation more complex. Rather separate Chapters are devoted for the OCR system of each language where there is an opportunity for the reader to select the best soft computing based OCR system for the corresponding language. Each language in Chaps. 4–9 is analyzed through 3 different soft computing OCR systems in a 2×2 matrix as shown in Fig. 1.1. It is to be noted that similar OCR systems are not used for the languages considered in the Chaps. 4–9. The different OCR systems used in this

Soft Computing based:	Chapter 4 English	Chapter 5 French	Chapter 6 German	Chapter 7 Latin	Chapter 8 Hindi	Chapter 9 Gujrati
OCR system 1	√					
OCR system 2	√	√	√	√	√	√
OCR system 3	√	√	√	√	√	√
OCR system 4		√	√	√		
OCR system 5					√	√

Fig. 1.1 The chapter wise distribution of the OCR systems in the monograph. (OCR system 1: Fuzzy Multilayer Perceptron [8]; OCR system 2: Rough Fuzzy Multilayer Perceptron [9, 10, 20]; OCR system 3: Fuzzy and Fuzzy Rough Support Vector Machines [5, 6, 7]; OCR system 4: Hierarchical Fuzzy Bidirectional Recurrent Neural Networks [4]; OCR system 5: Fuzzy Markov Random Fields [28])

monograph are highlighted in Fig. 1.1. Further insights into the OCR systems are available in Chap. 3. This will also help the reader to make a comparative analysis of each of the OCR system with respect to the six different languages considered.

An elementary knowledge of fuzzy and rough set theories as well as artificial neural networks [11] will also be helpful in understanding different concepts discussed in Chaps. 4–9. Several interesting soft computing tools [26] like neuro-fuzzy-genetic [12, 13] rough-neuro-fuzzy-genetic [20] etc. for OCR systems are discussed in these Chapters. The interested reader can refer the books [11, 17, 21, 30] on fuzzy sets, artificial neural networks, rough sets and genetic algorithms for better understanding.

The simulation results on the languages considered are discussed explicitly as separate sections in Chaps. 4–9. All the estimates and quantitative measures are borrowed from [8, 24]. The experimental results are based on the concepts given in Chaps. 2 and 3 as well as those highlighted in the specific Chapters. All the experiments are conducted on several real life OCR datasets using the MATLAB OCR toolbox [33, 34]. Several case studies are also illustrated in Chaps. 4–9. The entire computational framework is based on the basic concepts of soft computing. The interested reader can refer the books [19, 22] on soft computing. Thereafter, several accuracy measures are calculated with corresponding sensitivity analysis. The computational framework is implemented using MATLAB OCR toolbox [34]. A comparative analysis with other techniques is done for each individual language. The results for each individual language are evaluated through various assessment methods which are explained in Chaps. 4–9. All the assessment methods are adopted from [8] with experiments done using MATLAB OCR toolbox [34]. The summary and future research directions are given in Chap. 10.

Chapters 2 and 3 can be read independently. Chapters 4–9 cannot be read independently. They are based on the concepts illustrated in Chaps. 2 and 3 requires a better understanding of these chapters in order to understand the rest of the book. For example, in Chaps. 6 on OCR systems for german language can be well understood by the reader when he starts appreciating the soft computing tools for OCR in Chap. 3. The major prerequisite for better understanding of this book besides Chaps. 2 and 3 is basic knowledge of elementary mathematics [14]. We will cover the elementary mathematics as and when required with suitable pointers to different books [29].

1.2 Notation

It is very challenging to write a book containing appreciable amount of mathematical symbols and to achieve uniformity in the notation. Almost all the chapters of the book contain a good amount of mathematical analysis. We have tried to maintain a uniform notation within each chapter. This means that we may use the letters a and b to represent a closed interval $[a, b]$ in one chapter but they could stand for parameters in a kernel function in another chapter. We have used the following uniform notation throughout the book: (1) we place *tilde* over a letter to denote a fuzzy or rough set $\tilde{A}, \tilde{B}$ etc. and (2) all the fuzzy or rough sets are fuzzy or rough subsets of the real numbers. We use standard notations from mathematics and statistics as and when required.

The term *crisp* here denotes something which is neither fuzzy nor rough. A *crisp* set is a regular set and a *crisp* number is a real number. There is a potential problem with the symbol $\leq$. It usually means *fuzzy or rough subset* as $\tilde{A} \leq \tilde{B}$ which stands for $\tilde{A}$ is a fuzzy or rough subset of $\tilde{B}$. The meaning of the symbol $\leq$ should be clear from its use. Throughout the book $\bar{x}$ denotes mean of a random sample and not a fuzzy or rough set. It is explicitly pointed when this first arises in the book. Let $N(\mu, \sigma^2)$ denote the normal distribution with mean μ and variance σ^2. The critical values for the normal distribution are written as z_γ for hypothesis testing (confidence intervals). We have $\text{Prob}(X \geq z_\gamma) = \gamma$. Similarly the critical values of other distributions such as χ^2 distribution are also specified as and when required.

1.3 State of Art

The major part of this research falls in the intersection of OCR [23], feature extraction [16], character recognition languages [8], fuzzy sets [27], rough sets [21], artificial neural networks [11], soft computing [22], character recognition accuracy [8] and uncertainty analysis [8]. The references in Chap. 1 give the complete list of papers in these areas. The OCR researchers have their own research groups [31] which contains links to several basic papers in the area as well as journals in character recognition. There are several research papers available in the literature devoted to OCR and soft computing. The interested reader can always search these topics on his favourite search engine. Several papers on OCR employ second order probabilities, upper/lower probabilities etc. to model uncertainties. Here we use fuzzy and rough numbers [21, 27] to model uncertainty in the OCR datasets. We can also use crisp intervals to express the uncertainties. However, we do not use standard interval arithmetic to combine the uncertainties. We substitute fuzzy or rough numbers for uncertainty probabilities but we do not use fuzzy or rough probability theories to propagate uncertainty through the models. Most of our methods use fuzzy as well as rough numbers for expressing possibilities. A restricted amount of fuzzy and rough arithmetic is used to calculate fuzzy and rough probabilities. The statistical theory is based on probability theory. So fuzzy and rough statistics can take many forms depending on what probability (imprecise, interval, fuzzy, rough) theory is being used. Some key references in this exciting area are [1, 2] where the reader can discover many more other references.

1.4 Research Issues and Challenges

The research issues and challenges addressed during the entire course of this work
are [8]: (a) high dimensional massive datasets (b) over-fitting and assessing statistical significance (c) understanding the data patterns (d) nonstandard and incomplete data (e) mixed media data and integration. Further we also concentrated on
the scaling problem of different soft computing techniques to complex OCR datasets [8]. From the point of view of complexity analysis, for most scaling problems
the limiting factor of any OCR dataset has been number of examples and their
dimensions. A large number of examples introduces potential problems with both
time and space complexity. The time and space complexity aspects were addressed
through (a) data reduction (b) dimensionality reduction (c) active learning (d) data
partitioning and (e) efficient search algorithms.

1.5 Figures

All the figures and graphs in the book are created using different methods and
tools. The figures are adopted from standard texts of OCR [23] and soft computing [12]. The graphs are plotted based on standard datasets available from various
sources. Valid pointers are given for different datasets throughout the text book.
Some of the graphs are first plotted in Excel [32] and then exported to MS Word.
Likewise the other graphs are plotted in MATLAB [34] and then exported to MS
Word. The interested can refer the suggested references to graphs as per the needs
and requirements of the applications. All efforts are made to preserve the quality
of the graphs and figures towards better readability for the reader.

1.6 MATLAB OCR Toolbox

We have used the MATLAB OCR toolbox [34] to perform experiments with different soft computing techniques on six different languages viz english, french,
german, latin, hindi and gujrati languages. The experimental results are highlighted in Chaps. 4–9. Some of the OCR toolbox commands [33] are also used
to address different problems as well as towards the evaluation of the results
highlighted in the abovementioned Chapters. As an illustration consider the
MATLAB character classifier graphical user interface (GUI) shown in Fig. 1.2.
The MATLAB GUI encapsulates the steps involved with training an OCR system.
The GUI permits the user to load, binarize and segment images as well as compute
and plot features and save them for future analysis.

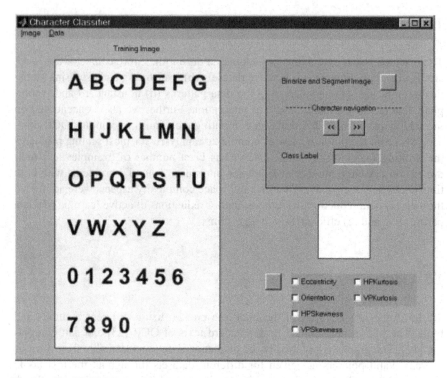

Fig. 1.2 The MATLAB character classifier graphical user interface

References

1. Berg, J. Van Den, Bergh, W. M. Van Den, Kaymak, U., Probabilistic and Statistical Fuzzy Set Foundations of Competitive Exception Learning, ERIM Report Series Research in Management, 2001.
2. Buckley, J. J., Fuzzy Probability and Statistics, Studies in Fuzziness and Soft Computing, Springer Verlag, 2006.
3. Bunke, H., Wang, P. S. P. (Editors), Handbook of Character Recognition and Document Image Analysis, World Scientific, 1997.
4. Chaudhuri, A., Ghosh, S. K., Sentiment Analysis of Customer Reviews Using Robust Hierarchical Bidirectional Recurrent Neural Network, Book Chapter: Artificial Intelligence Perspectives in Intelligent Systems, Radek Silhavy, Roman Senkerik, Zuzana Kominkova Oplatkova, Petr Silhavy, Zdenka Prokopova, (Editors), Advances in Intelligent Systems and Computing, Springer International Publishing, Switzerland, Volume 464, pp 249–261, 2016.
5. Chaudhuri, A., Fuzzy Rough Support Vector Machine for Data Classification, International Journal of Fuzzy System Applications, 5(2), pp 26–53, 2016.
6. Chaudhuri, A., Modified Fuzzy Support Vector Machine for Credit Approval Classification, AI Communications, 27(2), pp 189–211, 2014.
7. Chaudhuri, A., De, Fuzzy Support Vector Machine for Bankruptcy Prediction, Applied Soft Computing, 11(2), pp 2472–2486, 2011.
8. Chaudhuri, A., Some Experiments on Optical Character Recognition Systems for different Languages using Soft Computing Techniques, Technical Report, Birla Institute of Technology Mesra, Patna Campus, India, 2010.

9. Chaudhuri, A., De, K., Job Scheduling using Rough Fuzzy Multi-Layer Perception Networks, Journal of Artificial Intelligence: Theory and Applications, 1(1), pp 4–19, 2010.

10. Chaudhuri, A., De, K., Chatterjee, D., Discovering Stock Price Prediction Rules of Bombay Stock Exchange using Rough Fuzzy Multi-Layer Perception Networks, Book Chapter: Forecasting Financial Markets in India, Rudra P. Pradhan, Indian Institute of Technology Kharagpur, (Editor), Allied Publishers, India, pp 69–96, 2009.

11. Haykin, S., Neural Networks and Learning Machines, 3rd Edition, Prentice Hall, 2008.

12. Jang, J. S. R., Sun, C. T., Mizutani, E., Neuro-Fuzzy and Soft Computing: A Computational Approach to Learning and Machine Intelligence, Prentice Hall, 1997.

13. Kosko, B., Neural Networks and Fuzzy Systems: A Dynamical Systems Approach to Machine Intelligence, Prentice Hall of India, 2008.

14. Kreyszig, E., Advanced Engineering Mathematics, 10th Edition, Wiley International Press, 2010.

15. Leondes, C. T., Image Processing and Pattern Recognition, 1st Edition, Elsevier, 1997.

16. Liu, H, Motoda, H., Feature Extraction, Construction and Selection: A Data Mining Perspective, Kluwer Academic, 1998.

17. Mitchell, M., An Introduction to Genetic Algorithms, MIT Press, 1998.

18. Mollah, A. F., Majumder, N., Basu, S., Nasipuri, M., Design of an Optical Character Recognition System for Camera based Handheld Devices, International Journal of Computer Science Issues, 8 (4), pp 283–289, 2011.

19. Padhy, N. P., Simon, S. P., Soft Computing: With MATLAB Programming, Oxford University Press, 2015.

20. Pal, S. K, Mitra, S., Mitra, P., Rough Fuzzy MLP: Modular Evolution, Rule Generation and Evaluation, IEEE Transactions on Knowledge and Data Engineering, 15 (1), pp 14–25, 2003.

21. Polkowski, L, Rough Sets – Mathematical Foundations, Advances in Intelligent and Soft Computing, Springer Verlag, 2002.

22. Pratihar, D. K., Soft Computing, Alpha Science International Limited, 2007.

23. Rice, S. V., Nagy, G., Nartker, T. A., Optical Character Recognition: An Illustrated Guide to the Frontier, The Springer International Series in Engineering and Computer Science, Springer US, 1999.

24. Yamasaki, I., Quantitative Evaluation of Print Quality for Optical Character Recognition Systems, IEEE Transactions on Systems, Man and Cybernetics, 8 (5), pp 371–381, 1978.

25. Yu, F. T. S., Jutamulia, S. (Editors), Optical Pattern Recognition, Cambridge University Press, 1998.

26. Zadeh, L. A., Fuzzy Logic, Neural Networks and Soft Computing, Communications of the ACM, 37(3), pp 77–84, 1994.

27. Zadeh, L. A., Fuzzy Sets, Information and Control, 8(3), pp 338–353, 1965.

28. Zeng, J., Liu, Z. Q., Type-2 Fuzzy Markov Random Fields and their Application to Handwritten Chinese Character Recognition, IEEE Transactions on Fuzzy Systems, 16(3), pp 747–760, 2008.

29. Zill, D. G., Wright, W. S., Advanced Engineering Mathematics, 4th Edition, Jones and Bartlett Private Limited, 2011.

30. Zimmermann, H. J., Fuzzy Set Theory and its Applications, 4th Edition, Kluwer Academic Publishers, Boston, 2001.

31. https://tev-static.fbk.eu/OCR/ResearchProjects.html.

32. https://office.live.com/start/Excel.aspx.

33. http://in.mathworks.com/help/vision/optical-character-recognition-ocr.html.

34. http://in.mathworks.com/products/image/.

Chapter 2
Optical Character Recognition Systems

Abstract Optical character recognition (OCR) is process of classification of optical patterns contained in a digital image. The character recognition is achieved through segmentation, feature extraction and classification. This chapter presents the basic ideas of OCR needed for a better understanding of the book. The chapter starts with a brief background and history of OCR systems. Then the different techniques of OCR systems such as optical scanning, location segmentation, pre-processing, segmentation, representation, feature extraction, training and recognition and post-processing. The different applications of OCR systems are highlighted next followed by the current status of the OCR systems. Finally, the future of the OCR systems is presented.

Keywords OCR · Segmentation · Feature extraction · Classification

2.1 Introduction

Optical character recognition (OCR) [2, 7] is process of classification of optical patterns contained in a digital image corresponding to alphanumeric or other characters. The character recognition is achieved through important steps of segmentation, feature extraction and classification [12]. OCR has gained increasing attention in both academic research and in industry. In this chapter we have collected together the basic ideas of OCR needed for a better understanding of the book. It has been man's ancient dream to develop machines which replicate human functions. One such replication of human functions is reading of documents encompassing different forms of text. Over the last few decades machine reading has grown from dream to reality through the development of sophisticated and robust Optical character recognition (OCR) systems. OCR technology enables us to convert different types of documents such as scanned paper documents, pdf files or images captured by a digital camera into editable and searchable data. OCR systems have become one of the most successful applications of technology

© Springer International Publishing AG 2017

A. Chaudhuri et al., *Optical Character Recognition Systems for Different Languages with Soft Computing*, Studies in Fuzziness and Soft Computing 352, DOI 10.1007/978-3-319-50252-6_2

in pattern recognition and artificial intelligence fields. Though many commercial systems for performing OCR exist for a wide variety of applications, the available machines are still not able to compete with human reading capabilities with desired accuracy levels.

OCR belongs to the family of machine recognition techniques performing automatic identification. Automatic identification is the process where the recognition system identifies objects automatically, collects data about them and enters data directly into computer systems i.e. without human involvement. The external data is captured through analysis of images, sounds or videos. To capture data, a transducer is employed that converts the actual image or sound into a digital file. The file is then stored and at a later time it can be analyzed by the computer.

We start with a review of currently available automatic identification techniques and define OCR's position among them. The traditional way of entering data in a computer is through the keyboard. However, this is not always the best or the most efficient way. The automatic identification may serve as an alternative in many cases. There exist various techniques for automatic identification which cover the needs for different application areas. Some notable technologies and their applications worth mentioning apart from OCR are speech recognition, radio frequency, vision systems, magnetic stripe, bar code, magnetic ink and optical mark reading. These technologies have been actively used in past decades [10]. Here we introduce these technologies briefly from application point of view. Interested readers can refer [3, 5, 6, 8, 9, 11] for more elaborate discussion on these technologies:

(a) In speech recognition systems spoken input from a predefined library of words are recognized. Such systems are speaker independent and are generally used for reservations or telephonic ordering of goods. Another kind of such systems are those which are used to recognize speaker rather than words for identification.

(b) The radio frequency identification is often used in connection with toll roads for identification of cars. Special equipment on the car emits the information. The identification is efficient but special equipment is required both to send and to read the information. The information is inaccessible to humans.

(c) The vision systems are enforced through the usage TV camera where the objects are identified by their shape or size. This approach is generally used in automatons for recirculation of bottles. The type of bottle must be recognized first as the amount reimbursed for a bottle depends on its type.

(d) The information contained in magnetic stripes are widely used on credit cards etc. Quite a large amount of information can be stored on the magnetic stripe but specially designed readers are required and the information cannot be read by humans.

(e) The bar code consists of several dark and light lines representing a binary code for an eleven digit number, ten of which identify the particular product. The bar code is read optically when the product moves over glass window by a focused laser beam of weak intensity which is swept across glass window in a specially designed scanning pattern. The reflected light is measured and analysed by computer. Due to early standardization bar codes are today widely used and constitute a major share of the total market for automatic identification. The bar code represents a unique number that identifies the product and a price lookup is necessary to retrieve information about the price. The binary pattern representing the barcode takes up much space considering the small amount of information it actually contains. The barcodes are not readable to humans. Hence, they are only useful when the information is printed elsewhere in a human readable form or when human readability is not required.

(f) The printing in magnetic ink is mainly used within bank applications. The characters are written in ink that contains finely ground magnetic material. They are written in stylized fonts which are specifically designed for the application. Before the characters are read the ink is exposed to a magnetic field. This process accentuates each character and helps simplify the detection. The characters are read by interpreting the waveform obtained when scanning the characters horizontally. Each character is designed to have its own specific waveform. Although designed for machine reading, the characters are still readable to humans. However, reading is dependent on characters being printed with magnetic ink.

(g) The optical mark reading technology is used to register location of marks. It is used to read forms where the information is given by marking predefined alternatives. Such forms are also readable to humans. This approach is efficient when input is constrained. It is predefined with fixed number of alternatives.

OCR tries to address several issues of abovementioned techniques for automatic identification. They are required when the information is readable both to humans and machines. OCR systems have carved a niche place in pattern recognition. Their uniqueness lies in the fact that it does not require control of process that produces information. OCR deals with the problem of recognizing optically processed characters. Optical recognition is performed offline after the writing or printing has been completed whereas the online recognition is achieved where computer recognizes the characters as they are drawn. Both hand printed and printed characters may be recognized but the performance is directly dependent upon the quality of input documents. The more constrained the input is, better is the performance of OCR system. But when it comes to totally unconstrained handwriting performance of OCR machines is still questionable. The Fig. 2.1 shows the schematic representation of different areas of character recognition.

This chapter is organized as follows. A brief historical background of OCR systems is Sect. 2.2. In Sect. 2.3 a discussion of different techniques of OCR is highlighted. This is followed by the applications of OCR systems in Sect. 2.4. In Sect. 2.5 we present the status of the OCR systems. Finally in Sect. 2.6 the future of OCR systems is given.

Fig. 2.1 The different areas
of character recognition

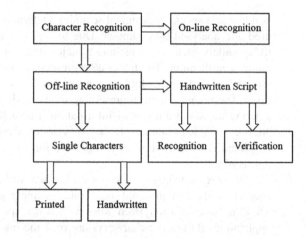

2.2 Optical Character Recognition Systems: Background and History

Character recognition is a subset of pattern recognition area. Several concepts and techniques in OCR are borrowed from pattern recognition and image processing. However, it was character recognition that provided impetus for making pattern recognition and image analysis as matured fields of science and engineering.

Writing which has been the most natural mode of collecting, storing and transmitting information through the centuries now serves not only for communication among humans but also serves for communication of humans and machines. The intensive research effort in the field of OCR was not only because of its challenge on simulation of human reading but also because it provides efficient applications such as the automatic processing of bulk amount of papers, transferring data into machines and web interface to paper documents. To replicate human functions by machines and making the machine perform common tasks like reading is an ancient dream. The origins of character recognition dates back to 1870 when C.R. Carey of Boston Massachusetts [3, 7, 8] invented retina scanner which was an image transmission system using a mosaic of photocells. Early versions needed to be trained with images of each character and worked on one font at a time.

The history of OCR can be traced as early as 1900, when the Russian scientist Tyuring attempted to develop an aid for the visually handicapped [6]. The first character recognizers appeared in the middle of the 1940s with the development of digital computers [3]. The early work on the automatic recognition of characters has been concentrated either upon machine printed text or upon a small set of well distinguished handwritten text or symbols. Machine printed OCR systems in this period generally used template matching in which an image is compared to a library of images. For handwritten text, low level image processing techniques have been used on the binary image to extract feature vectors which are then fed to statistical classifiers. Successful but constrained algorithms have been implemented mostly for Latin characters and numerals. However, some studies on Japanese, Chinese, Hebrew, Indian, Cyrillic, Greek, and Arabic characters and numerals in both machine-printed and handwritten cases were also initiated [3].

Two decades later Nipkow [11] invented sequential scanner which was a major breakthrough both for modern television and reading machines. During the first few decades of 19th century several attempts were made to develop devices to aid the blind through experiments with OCR [8]. However, the modern version of OCR did not appear till the mid 1940s when digital computer came into force. The motivation for development of OCR systems started from then onwards when people thought for possible business and commercial applications.

By 1950 the technological revolution [3, 6] was moving forward at high speed and electronic data processing was becoming an upcoming and important field. The commercial character recognizers available in 1950s where electronic tablets captured the x–y coordinate data of pen tip movement was first introduced. This innovation enabled the researchers to work on the online handwriting recognition problem [3]. The data entry was performed through punched cards. A cost effective way of handling the increasing amount of data was then required. At the same time the technology for machine reading was becoming sufficiently mature for application. By mid 1950s OCR machines became commercially available [8]. The first OCR reading machine [9] was installed at Reader's Digest in 1954. This equipment was used to convert typewritten sales reports into punched cards for input into the computer.

The commercial OCR systems appearing from 1960 to 1965 were often referred to as first generation OCR [3, 8]. The OCR machines of this generation were mainly characterized by constrained letter shapes. The symbols were specially designed for machine reading. When multi-font machines started to appear, they could read up to several different fonts. The number of fonts were limited by pattern recognition method applied and template matching which compares the character image with library of prototype images for each character of each font.

In mid 1960s and early 1970s the reading machines of second generation appeared [3, 8]. These systems were able to recognize regular machine printed characters and also had hand printed character recognition capabilities. When hand printed characters were considered, the character set was constrained to numerals as well as few letters and symbols. The first and famous system of this kind was IBM 1287 in 1965. During this period Toshiba developed the first automatic letter sorting machine for postal code numbers. Hitachi also made the first OCR machine for high performance and low cost. In this period significant work was done in the area of standardization. In 1966 a thorough study of OCR requirements was completed and an American standard OCR character set was defined as OCR–A shown in Fig. 2.2. This font was highly stylized and designed to facilitate optical recognition although still readable to humans. A European font was also designed as OCR–B shown in Fig. 2.3 which had more natural fonts than American standard. Attempts were made to merge two fonts in one standard through machines which could read both standards.

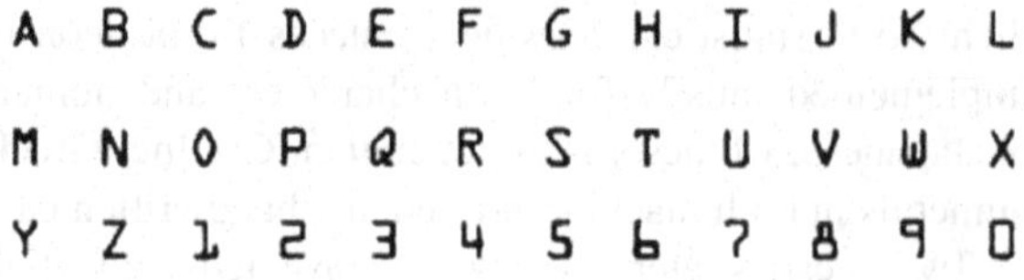

Fig. 2.2 OCR–A font   A B C D E F G H I J K L

                             M N O P Q R S T U V W X

                             Y Z 1 2 3 4 5 6 7 8 9 0

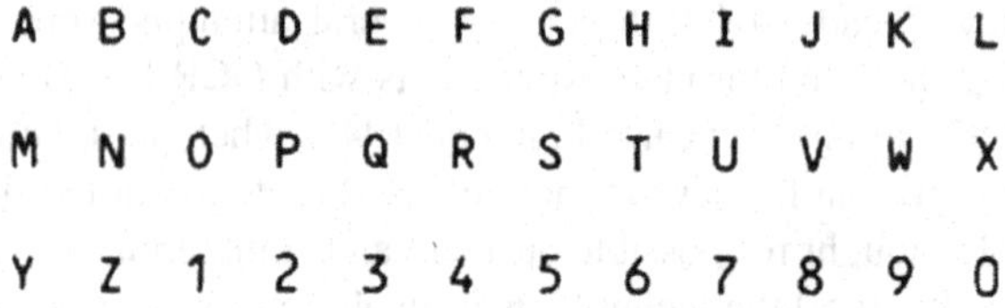

Fig. 2.3 OCR–B font   A B C D E F G H I J K L

                             M N O P Q R S T U V W X

                             Y Z 1 2 3 4 5 6 7 8 9 0

In mid 1970s the third generation of OCR systems appeared [3, 8]. The challenge was handing documents of poor quality and large printed and hand written character sets. The low cost and high performance objectives were achieved through dramatic advances in hardware technology. This resulted in the growth of sophisticated OCR machines for users. In the period before personal computers and laser printers started to dominate the area of text production, typing was a special niche for OCR. The uniform print spacing and small number of fonts made simply designed OCR devices very useful. Rough drafts could be created on ordinary typewriters and fed into computer through an OCR device for final editing. In this way word processors which were an expensive resource at this time could support several people at reduced equipment costs.

Although OCR machines became commercially available already in the 1950s, only few thousand systems were sold till 1986 [6]. The main reason for this was the cost of systems. However, as hardware prices went down and OCR systems started to become available as software packages, the sale increased considerably. Advanced systems capable of producing a high degree of recognition accuracy for most fonts are now common. Some systems are capable of reproducing formatted output that closely approximates the original page including images, columns, and other non-textual components. Today few millions of OCR systems are sold every week. The cost of omnifont OCR has dropped with a factor of ten every other year for the last few decades.

The studies up until 1980 suffered from the lack of powerful computer hardware and data acquisition devices. With the explosion of information technology, the previously developed methodologies found a very fertile environment for rapid growth in many application areas, as well as OCR system development [3]. The structural approaches were initiated in many systems in addition to the statistical methods [3]. The OCR research was focused basically on the shape recognition techniques without using any semantic information. This led to an upper limit in the recognition rate which was not sufficient in many practical applications.

The real progress on OCR systems achieved during 1990s using the new development tools and methodologies which are empowered by the continuously growing information technologies. In the early 1990s, image processing and pattern recognition techniques were efficiently combined with artificial intelligence methodologies. Researchers developed complex OCR algorithms, which receive high-resolution input data and require extensive number crunching in the implementation phase. Nowadays, in addition to the more powerful computers and more accurate electronic equipments such as scanners, cameras, and electronic tablets, we have efficient, modern use of methodologies such as artificial neural networks (ANNs), hidden Markov models (HMMs), fuzzy set reasoning and natural language processing. The recent systems for the machine printed offline [1, 3] and limited vocabulary, user dependent online handwritten characters [1] are quite satisfactory for restricted applications. However, still a long way to go in order to reach the ultimate goal of machine simulation of fluent human reading especially for unconstrained online and offline handwriting.

2.3 Techniques of Optical Character Recognition Systems

The main concept in automatic recognition of patterns is first to teach the machine which class of patterns that may occur and what they look like [3, 4]. In OCR patterns are letters, numbers and some special symbols like commas, question marks as well as different characters. The teaching of machine is performed by showing machine examples of characters of all different classes. Based on these examples the machine builds prototype or description of each class of characters. During recognition the unknown characters are compared to previously obtained descriptions and assigned to class that gives the best match. In most commercial systems for character recognition training process is performed in advance. Some systems however include facilities for training in the case of inclusion of new classes of characters.

A typical OCR system consists of several components as shown in Fig. 2.4 [3, 7]. The first step is to digitize analog document using an optical scanner. When regions containing text are located each symbol is extracted through segmentation process. The extracted symbols are pre-processed, eliminating noise to facilitate feature extraction. The identity of each symbol is found by comparing extracted features with descriptions of symbol classes obtained through a previous learning phase. Finally contextual information is used to reconstruct words and numbers of the original text. These steps are briefly presented here. Interested readers can refer [11] for more elaborate discussion of OCR system components.

2.3.1 Optical Scanning

The first component in OCR is optical scanning. Through scanning process digital image of original document is captured. In OCR optical scanners are used which

Fig. 2.4 The components of
an OCR system

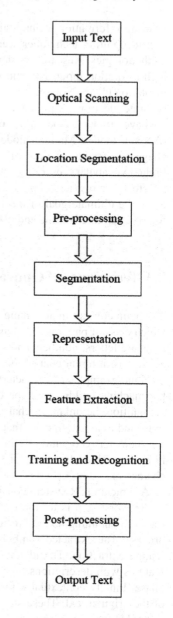

consist of transport mechanism and sensing device that converts light intensity into grey levels. Printed documents consist of black print on white background. When performing OCR multilevel image is converted into bi-level black and white image. This process known as thresholding is performed on scanner to save memory space and computational effort. The thresholding process is important as the results of recognition are totally dependent on quality of bi-level image. A fixed threshold is used where gray levels below this threshold are black and levels above are white. For high contrast document with uniform background a pre-chosen fixed threshold can be sufficient. However, documents encountered in practice have rather large range. In these cases more sophisticated methods for thresholding are required to obtain good results. The best thresholding methods vary threshold adapting to local properties of document such as contrast and brightness. However, such methods usually depend on multilevel scanning of document which requires more memory and computational capacity.

2.3.2 Location Segmentation

The next OCR component is location segmentation. Segmentation determines constituents of an image. It is necessary to locate regions of document which have printed data and are distinguished from figures and graphics. For example, when performing automatic mail sorting through envelopes address must be located and separated from other prints like stamps and company logos, prior to recognition. When applied to text, segmentation is isolation of characters or words. Most of OCR algorithms segment words into isolated characters which are recognized individually. Usually segmentation is performed by isolating each connected component. This technique is easy to implement but problems arise if characters touch or they are fragmented and consist of several parts. The main problems in segmentation are: (a) extraction of touching and fragmented characters (b) distinguishing noise from text (c) misinterpreting graphics and geometry with text and vice versa. For interested readers further details are available in [11].

2.3.3 Pre-processing

The third OCR component is pre-processing. The raw data depending on the data acquisition type is subjected to a number of preliminary processing steps to make it usable in the descriptive stages of character analysis. The image resulting from scanning process may contain certain amount of noise. Depending on the scanner resolution and the inherent thresholding, the characters may be smeared or broken. Some of these defects which may cause poor recognition rates and are eliminated through pre-processor by smoothing digitized characters. Smoothing implies both filling and thinning. Filling eliminates small breaks, gaps and holes in digitized characters while thinning reduces width of line. The most common technique for smoothing moves a window across binary image of character and applies certain rules to the contents of window. Pre-processing also includes normalization alongwith smoothing. The normalization is applied to obtain characters of uniform size, slant and rotation. The correct rotation is found through its angle. For rotated pages and lines of text, variants of Hough transform are commonly used for detecting skew.

The pre-processing component thus aims to produce data that are easy for the OCR systems to operate accurately. It is an important activity to be performed before the actual data analysis. The main objectives of pre-processing can be pointed as [1, 3]: (a) noise reduction (b) normalization of the data and (c) compression in the amount of information to be retained. In rest of this subsection the aforementioned objectives of pre-processing objectives are discussed with the corresponding techniques.

(a) Noise reduction: The noise introduced by the optical scanning device or the writing instrument causes disconnected line segments, bumps and gaps in lines, filled loops, etc. The distortion including local variations, rounding of corners, dilation and erosion is a potential problem. It is necessary to eliminate these imperfections prior to actual processing of the data. The noise reduction techniques can be categorized in three major groups [1, 3]: (i) filtering (ii) morphological operations and (iii) noise modeling.

   (i)   Filtering aims to remove noise and diminish spurious points usually introduced by uneven writing surface and poor sampling rate of the data acquisition device. Various spatial and frequency domain filters have been designed for this purpose. The basic idea is to convolute a predefined mask with the image to assign a value to a pixel as a function of the gray values of its neighboring pixels. Several filters have been designed for smoothing, sharpening, thresholding, removing slightly textured or coloured background and contrast adjustment purposes [1, 3].

   (ii) The basic idea behind the morphological operations is to filter the character image replacing the convolution operation by the logical operations. Various morphological operations have been designed to connect the broken strokes, decompose the connected strokes, smooth the contours, prune wild points, thin the characters and extract the boundaries [1, 3]. The morphological operations can be successfully used to remove noise on the character images due to low quality of paper and ink as well as erratic hand movement.

   (iii) Noise can generally be removed by calibration techniques if it would have been possible to model it. However, noise modeling is not possible in most of the applications. There exists some available literature on noise modeling introduced by optical distortion such as speckle, skew and blur. It is also possible to assess the quality of the character images and remove the noise to a certain degree [1, 3].

(b) Normalization: The normalization methods aim to remove the variations of the writing and obtain standardized data. Some of the commonly used methods for normalization are [1, 3]: (i) skew normalization and baseline extraction (ii) slant normalization (iii) size normalization and (iv) contour smoothing.

   (i)   Skew normalization and baseline extraction: Due to inaccuracies in the scanning process and writing style the writing may be slightly tilted or curved within the image. This can hurt the effectiveness of the algorithms and thus should be detected and corrected. Additionally, some

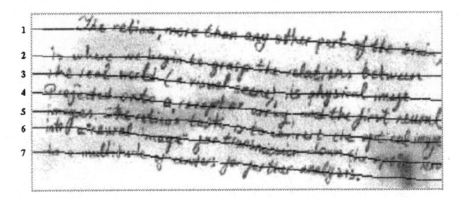

Fig. 2.5 The baseline extraction using attractive and repulsive network

characters are distinguished according to the relative position with respect to the baseline, such as 9 and g. The methods of baseline extraction include using the projection profile of the image, nearest neighbor clustering, cross correlation method between lines and Hough transform [1, 3]. An attractive repulsive nearest neighbor is used for extracting the baseline of complicated handwriting in heavy noise [1, 3] as shown in Fig. 2.5. After skew detection the character or word is translated to the origin, rotated or stretched until the baseline is horizontal and retranslated back into the display screen space.

(ii) Slant normalization: One of the measurable factors of different handwriting styles is the slant angle between longest stroke in a word and the vertical direction. Slant normalization is used to normalize all characters to a standard form. The most common method for slant estimation is the calculation of the average angle of near vertical elements as shown in Fig. 2.6a, b. The vertical line elements from contours are extracted by tracing chain code components using a pair of one dimensional filters [1, 3]. The coordinates of the start and end points of each line element provide the slant angle. The projection profiles are computed for a number of angles away from the vertical direction [1, 3]. The angle corresponding to the projection with the greatest positive derivative is used to detect the least amount of overlap between vertical strokes and the dominant slant angle. The slant detection is performed by dividing the image into vertical and horizontal windows [1, 3]. The slant is estimated based on the center of gravity of the upper and lower half of each window averaged over all the windows. A variant of the Hough transform is used by scanning left to right across the image and calculating projections in the direction of 21 different slants [1, 3]. The top three projections for any slant are added and the slant with the largest count is taken as the slant value. In some cases the recognition systems do not use slant correction and compensate it during training stage [1, 3].

(a)

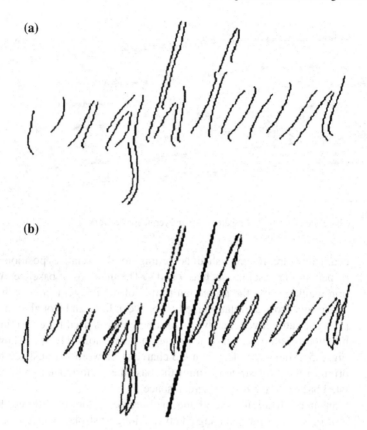

(b)

Fig. 2.6 **a** Slant angle estimation near vertical elements. **b** Slant angle estimation average slant angle

(iii) Size Normalization is used to adjust the character size to a certain standard. The OCR methods may apply for both horizontal and vertical size normalizations. The character is divided into number of zones and each of these zones is separately scaled [1, 3]. The size normalization can also be performed as a part of the training stage and the size parameters are estimated separately for each particular training data [1, 3]. In Fig. 2.7 two sample characters are gradually shrunk to the optimal size which maximize the recognition rate in the training data. The word recognition preserves large intra class differences in the length of words so they may also assist in recognition; it tends to only involve vertical height normalization or bases the horizontal size normalization on the scale factor calculated for vertical normalization [1, 3].

(iv) Contour smoothing eliminates the errors due to the erratic hand motion during the writing. It generally reduces the number of sample points needed to represent the script and thus improves efficiency in remaining pre-processing steps [1, 3].

Fig. 2.7 The normalization
of characters

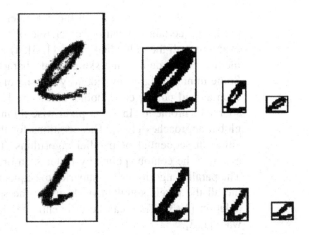

(c) Compression: It is well known that classical image compression techniques
transform the image from the space domain to domains which are not suitable for recognition. The compression for OCR requires space domain techniques for preserving the shape information. The two popular compression
techniques used are: (i) thresholding and (ii) thinning.

   (i)   Thresholding: In order to reduce storage requirements and to increase
         processing speed it is often desirable to represent gray scale or color
         images as binary images by picking a threshold value. The two impor-
         tant categories of thresholding are viz global and local. The global
         thresholding picks one threshold value for the entire character image
         which is often based on an estimation of the background level from the
         intensity histogram of the image [1, 3]. The local or adaptive threshold-
         ing use different values for each pixel according to the local area infor-
         mation [1, 3]. A comparison of common global and local thresholding
         techniques is given by using an evaluation criterion that is goal directed
         keeping in view of the desired accuracy of the OCR system [1, 3]. It
         has been shown that Niblack's locally adaptive method [1, 3] produces
         the best result. An adaptive logical method is developed [1, 3] by ana-
         lyzing the clustering and connection characteristics of the characters in
         degraded images.

   (ii) Thinning: While it provides a tremendous reduction in data size, thin-
         ning extracts the shape information of the characters. Thinning can be
         considered as conversion of offline handwriting to almost online like
         data with spurious branches and artifacts. The two basic approaches for
         thinning are based on pixel wise and non-pixel wise thinning [1, 3]. The
         pixel wise thinning methods locally and iteratively process the image
         until one pixel wide skeleton remains. They are very sensitive to noise
         and deforms the shape of the character. The non-pixel wise methods use

some global information about the character during the thinning. They produce a certain median or center line of the pattern directly without examining all the individual pixels [1, 3]. The clustering based thinning method [1, 3] defines the skeleton of character as the cluster centers. Some thinning algorithms identify the singular points of the characters such as end points, cross points and loops [1, 3]. These points are the source of problems. In a non-pixel wise thinning they are handled with global approaches [1, 3]. The iterations for thinning can be performed either in sequential or parallel algorithms. The sequential algorithms examine the contour points by raster scan or contour following [1, 3]. The parallel algorithms are superior to sequential ones since they examine all the pixels simultaneously using the same set of conditions for deletion [1, 3]. They can be efficiently implemented in parallel hardware [1, 3].

It is to be noted that the above techniques affect the data and may introduce unexpected distortions to the character image. As a result these techniques may cause the loss of important information about writing and thus should be applied with care.

2.3.4 Segmentation

The pre-processing stage yields a clean character image in the sense that a sufficient amount of shape information, high compression, and low noise on a normalized image is obtained. The next OCR component is segmentation. Here the character image is segmented into its subcomponents. Segmentation is important because the extent one can reach in separation of the various lines in the characters directly affects the recognition rate. Internal segmentation is used here which isolates lines and curves in the cursively written characters. Though several remarkable methods have developed in the past and a variety of techniques have emerged, the segmentation of cursive characters is an unsolved problem. The character segmentation strategies are divided into three categories [1, 3]: (a) explicit segmentation (b) implicit segmentation and (c) mixed strategies.

(a) In explicit segmentation the segments are identified based on character like properties. The process of cutting up the character image into meaningful components is achieved through dissection. Dissection analyzes the character image without using a specific class of shape information. The criterion for good segmentation is the agreement of general properties of the segments with those expected for valid characters. The available methods based on the dissection of the character image use white space and pitch, vertical projection analysis, connected component analysis and landmarks. The explicit segmentation can be subjected to evaluation using the linguistic context [1, 3].

(b) The implicit segmentation strategy is based on recognition. It searches the image for components that matches the predefined classes. The segmentation is performed by using the recognition confidence including syntactic or semantic correctness of the overall result. In this approach two classes of methods are employed viz (i) methods that make some search process and (ii) methods that segment a feature representation of the image [1, 3]. The first class attempts to segment characters into units without use of feature based dissection algorithms. The image is divided systematically into many overlapping pieces without regard to content. These methods originate from schemes developed for the recognition of machine printed words [1, 3]. The basic principle is to use a mobile window of variable width to provide sequences of tentative segmentations which are confirmed by OCR. The second class of methods segments the image implicitly by classification of subsets of spatial features collected from the image as a whole. This can be done either through hidden markov chains or non markov based approaches. The non markov approach stem from the concepts used in machine vision for recognition of occluded object [1, 3]. This recognition based approach uses probabilistic relaxation, the concept of regularities and singularities and backward matching [1, 3].

(c) The mixed strategies combine explicit and implicit segmentation in a hybrid way. A dissection algorithm is applied to the character image, but the intent is to over segment i.e. to cut the image in sufficiently many places such that the correct segmentation boundaries are included among the cuts made. Once this is assured, the optimal segmentation is sought by evaluation of subsets of the cuts made. Each subset implies a segmentation hypothesis and classification is brought to bear to evaluate the different hypothesis and choose the most promising segmentation [1, 3]. The segmentation problem is formulated [1, 3] as finding the shortest path of a graph formed by binary and gray level document image. The hidden markov chain probabilities obtained from the characters of a dissection algorithm are used to form a graph [1, 3]. The optimum path of this graph improves the result of the segmentation by dissection and hidden markov chain recognition. The mixed strategies yield better results compared to explicit and implicit segmentation methods. The error detection and correction mechanisms are often embedded into the systems. The wise usage of context and classifier confidence generally leads to improved accuracy [1, 3].

2.3.5 Representation

The fifth OCR component is representation. The image representation plays one of the most important roles in any recognition system. In the simplest case, gray level or binary images are fed to a recognizer. However, in most of the recognition systems in order to avoid extra complexity and to increase the accuracy of the algorithms, a more compact and characteristic representation is required. For this purpose, a set of features is extracted for each class that helps distinguish it from other classes while remaining invariant to characteristic differences within the class [1, 3]. The character image representation methods are generally categorized into three major groups: (a) global transformation and series expansion (b) statistical representation and (c) geometrical and topological representation.

(a) Global transformation and series expansion: A continuous signal generally contains more information than needs to be represented for the purpose of classification. This may be true for discrete approximations of continuous signals as well. One way to represent a signal is by a linear combination of a series of simpler well defined functions. The coefficients of the linear combination provide a compact encoding known as transformation or series expansion. Deformations like translation and rotations are invariant under global transformation and series expansion. Some common transform and series expansion methods used in OCR are: (i) fourier transform (ii) gabor transform (iii) wavelets (iv) moments and (v) karhunen loeve expansion.

(i)   Fourier transform: The general procedure is to choose magnitude spectrum of the measurement vector as the features in an n-dimensional euclidean space. One of the most attractive properties of the fourier transform is the ability to recognize the position shifted characters when it observes the magnitude spectrum and ignores the phase.

(ii) Gabor transform: This is a variation of the windowed fourier transform. In this case, the window used is not a discrete size but is defined by a gaussian function [1, 3].

(iii) Wavelet transformation is a series expansion technique that allows us to represent the signal at different levels of resolution. The segments of character image correspond to the units of the character and are represented by wavelet coefficients corresponding to various levels of resolution. These coefficients are then fed to a classifier for recognition [1, 3]. The representation in multi resolution analysis with low resolution absorbs the local variation in handwriting as opposed to the high resolution. However, the representation in low resolution may cause the important details for the recognition stage to be lost.

(iv) Moments such as central moments, legendre moments and zernike moments form a compact representation of the original character image that make the process of recognizing an object scale, translation and rotation invariant [1, 3]. The moments are considered as series expansion representation since the original character image can be completely reconstructed from the moment coefficients.

(v)   Karhunen loeve expansion is an eigenvector analysis which attempts to reduce the dimension of the feature set by creating new features that are linear combinations of the original ones. It is the only optimal transform in terms of information compression. Karhunen loeve expansion is used in several pattern recognition problems such as face recognition.

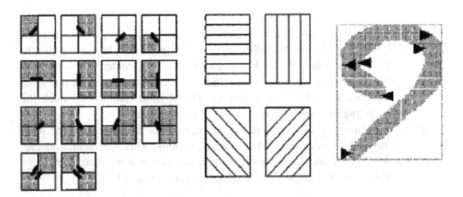

Fig. 2.8 The contour direction and bending point features with zoning

It is also used in the National Institute of Standards and Technology (NIST) OCR system for form based handprint recognition [1, 3]. Since it requires computationally complex algorithms, the use of karhunen loeve features in OCR problems is not widespread. However, by the continuous increase of the computational power, it has become a realistic feature for the current OCR systems [1, 3].

(b) Statistical representation: The representation of a character image by statistical distribution of points takes care of style variations to some extent. Although this type of representation does not allow the reconstruction of the original image, it is used for reducing the dimension of the feature set providing high speed and low complexity. Some of the major statistical features used for character representation are: (i) zoning (ii) crossings and distances and (iii) projections.

    (i)   Zoning: The frame containing the character is divided into several overlapping or non-overlapping zones. The densities of the points or some features in different regions are analyzed and form the representation. For example, contour direction features measure the direction of the contour of the character [1, 3] which are generated by dividing the image array into rectangular and diagonal zones and computing histograms of chain codes in these zones. Another example is the bending point features which represent high curvature points, terminal points and fork points [1, 3]. The Fig. 2.8 indicates contour direction and bending point features.

    (ii)   Crossings and distances: A popular statistical feature is the number of crossing of a contour by a line segment in a specified direction. The character frame is partitioned into a set of regions in various directions and then the black runs in each region are coded by the powers of two [1, 3]. Another study encodes the location and number of transitions from background to foreground pixels along vertical lines through the word [1, 3]. Also, the distance of line segments from a given boundary such as the upper and lower portion of the frame, they can be used as statistical features [1, 3]. These features imply that a horizontal threshold is established above, below and through the center of the normalized character. The number of times the character crosses a threshold becomes the value of that feature. The obvious intent is to catch the ascending and descending portions of the character.

(iii) Projections: The characters can be represented by projecting the pixel gray values onto lines in various directions. This representation creates a one dimensional signal from a two dimensional character image which can be used to represent the character image [1, 3].

(c) Geometrical and topological representation: The various global and local properties of characters can be represented by geometrical and topological features with high tolerance to distortions and style variations. This type of representation may also encode some knowledge about the structure of the object or may provide some knowledge as to what sort of components make up that object. The topological and geometrical representations can be grouped into: (i) extracting and counting topological structures (ii) measuring and approximating the geometrical properties (iii) coding and (iv) graphs and trees.

(i) Extracting and counting topological structures: In this representation group, a predefined structure is searched in a character. The number or relative position of these structures within the character forms a descriptive representation. The common primitive structures are the strokes which make up a character. These primitives can be as simple as lines and arcs which are the main strokes of Latin characters and can be as complex as curves and splines making up Arabic or Chinese characters. In online OCR, a stroke is also defined as a line segment from pen down to pen up [1, 3]. The characters can be successfully represented by extracting and counting many topological features such as the extreme points, maxima and minima, cusps above and below a threshold, openings to the right, left, up, and down, cross points, branch points, line ends, loops, direction of a stroke from a special point, inflection between two points, isolated dots, a bend between two points, symmetry of character, horizontal curves at top or bottom, straight strokes between two points, ascending, descending, and middle strokes and relations among the stroke that make up a character [1, 3]. The Fig. 2.9 indicates some of the topological features.

(ii) Measuring and approximating the geometrical properties: In many studies [1, 3] the characters are represented by the measurement of the geometrical quantities such as the ratio between width and height of the bounding box of a character, the relative distance between the last point and the last y-min, the relative horizontal and vertical distances between first and last points, distance between two points, comparative lengths

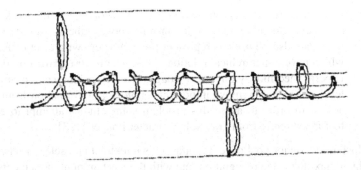

Fig. 2.9 The topological features: Maxima and minima on the exterior and interior contours, reference lines, ascenders and descenders

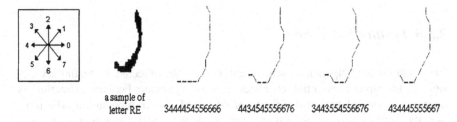

a sample of
letter RE   3444454556666   4434545556676   3443554556676   434445555667

Fig. 2.10   Sample arabic character and the chain codes of its skeleton

between two strokes and width of a stroke. A very important character-istic measure is the curvature or change in the curvature [1, 3]. Among many methods for measuring the curvature information, the suggestion measures local stroke direction distribution for directional decomposi-tion of the character image. The measured geometrical quantities can be approximated by a more convenient and compact geometrical set of features. A class of methods includes polygonal approximation of a thinned character [1, 3]. A more precise and expensive version of the polygonal approximation is the cubic spline representation [1, 3].

(iii) Coding: One of the most popular coding schema is freeman's chain code. This coding is essentially obtained by mapping the strokes of a character into a two dimensional parameter space which is made up of codes as shown in Fig. 2.10. There are many versions of chain coding. As an example, the character frame is divided to left right sliding win-dow and each region is coded by the chain code.

(iv) Graphs and trees: The characters are first partitioned into a set of topo-logical primitives such as strokes, loops, cross points etc. Then these primitives are represented using attributed or relational graphs [1, 3]. There are two kinds of image representation by graphs. The first kind uses the coordinates of the character shape [1, 3]. The second kind is an abstract representation with nodes corresponding to the strokes and edges corresponding to the relationships between the strokes [1, 3]. The trees can also be used to represent the characters with a set of features which have a hierarchical relation [1, 3]. The feature extraction process is performed mostly on binary images. However, binarization of a gray level image may remove important topological information from characters. In order to avoid this problem some studies attempt to extract features directly from grayscale character images [1, 3].

In conclusion, the major goal of representation is to extract and select a set of features which maximizes the recognition rate with the least amount of elements. The feature extraction and selection is defined [1, 3] as extracting the most representative information from the raw data which minimizes the within class pattern variability while enhancing the between class pattern variability.

2.3.6 Feature Extraction

The sixth OCR component is feature extraction. The objective of feature extraction is to capture essential characteristics of symbols. Feature extraction is accepted as one of the most difficult problems of pattern recognition. The most straight forward way of describing character is by actual raster image. Another approach is to extract certain features that characterize symbols but leaves the unimportant attributes. The techniques for extraction of such features are divided into three groups' viz. (a) distribution of points (b) transformations and series expansions and (c) structural analysis. The different groups of features are evaluated according to their noise sensitivity, deformation, ease of implementation and use. The criteria used in this evaluation are: (a) robustness in terms of noise, distortions, style variation, translation and rotation and (b) practical usage in terms of recognition speed, implementation complexity and independence. Some of the commonly used feature extraction techniques are template matching and correlation, transformations, distribution of points and structural analysis. For interested readers further details are available in [11].

Another important task associated with feature extraction is classification. Classification is the process of identifying each character and assigning to it correct character class. The two important categories of classification approaches for OCR are decision theoretic and structural methods. In decision theoretic recognition character description is numerically represented in feature vector. There may also be pattern characteristics derived from physical structure of character which are not as easily quantified. Here relationship between the characteristics may important when deciding on class membership. For example, if we know that a character consists of one vertical and one horizontal stroke it may be either 'L' or 'T'. The relationship between two strokes is required to distinguish characters. The principal approaches to decision theoretic recognition are minimum distance classifiers, statistical classifiers and neural networks. In structural recognition syntactic methods are the most prevalent approaches. A detailed discussion of these approaches is available in [3, 11].

2.3.7 Training and Recognition

The seventh OCR component is training and recognition. OCR systems extensively use the methodologies of pattern recognition which assigns an unknown sample into a predefined class. The OCR are investigated in four general approaches of pattern recognition as suggested in [1, 3]: (a) template matching (b) statistical techniques (c) structural techniques and (d) ANNs. These approaches are neither necessarily independent nor disjointed from each other. Occasionally, an OCR technique in one approach can also be considered to be a member of other approaches. In all of the above approaches, OCR techniques use either holistic or analytic strategies for the training and recognition stages. The holistic strategy employs top down approaches for recognizing the full character eliminating the segmentation problem. The price for this computational saving is to constrain the problem of OCR to limited vocabulary. Also, due to the complexity introduced by the representation of a single character or stroke the recognition accuracy is decreased. On the other hand, the analytic strategies employ bottom up approach starting from stroke or character level and going toward producing a meaningful text. The explicit or implicit segmentation algorithms are required for this strategy, not only adding extra complexity to the problem but also introducing segmentation error to the system. However, with the cooperation of segmentation stage, the problem is reduced to the recognition of simple isolated characters or strokes, which can be handled for unlimited vocabulary with high recognition rates.

(a) Template matching: The OCR techniques vary widely according to the feature set selected from the long list of features described in the previous section for image representation. The features can be as simple as the gray-level image frames with individual characters complicated as graph representation of character primitives. The simplest way of OCR is based on matching the stored prototypes against the character to be recognized. Generally speaking, matching operation determines the degree of similarity between two vectors such as group of pixels, shapes, curvature etc. in the feature space. The matching techniques can be classified in three classes: (i) direct matching (ii) deformable templates and elastic matching and (iii) relaxation matching.

(i) Direct matching: A gray level or binary input character is directly compared to a standard set of stored prototypes. According to the similarity measures such as euclidean, mahalanobis, jaccard or yule, a prototype matching is done for recognition. The matching techniques can be as simple as one-to-one comparison or as complex as decision tree analysis in which only selected pixels are tested. A template matcher can

Fig. 2.11 **a** The deformable
templates: deformations
of a sample digit. **b** The
deformable templates:
deformed template
superimposed on target image
with dissimilarity measures

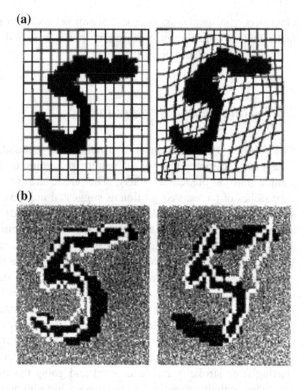

combine multiple information sources including match strength and
k-nearest neighbor measurements from different metrics [1, 3]. Although
direct matching method is intuitive and has a solid mathematical background, the recognition rate of this method is very sensitive to noise.

(ii)   Deformable templates and elastic matching: An alternative method is
the use of deformable templates where an image deformation is used to
match an unknown image against a database of known images. The two
characters are matched by deforming the contour of one to fit the edge
strengths of the other [1, 3]. A dissimilarity measure is derived from the
amount of deformation needed, the goodness of fit of the edges and the
interior overlap between the deformed shapes as shown in Fig. 2.11a, b.
The basic idea of elastic matching is to optimally match the unknown
symbol against all possible elastic stretching and compression of each
prototype. Once the feature space is formed, the unknown vector is
matched using dynamic programming and a warping function [1, 3].
Since the curves obtained from the skeletonization of the characters
could be distorted, elastic-matching methods cannot deal with topological correlation between two patterns in the OCR. In order to avoid this
difficulty, a self-organization matching approach is proposed in [1, 3]
for hand printed OCR using thick strokes. The elastic matching is also
popular in on-line recognition systems [1, 3].

(iii) Relaxation matching: It is a symbolic level image matching technique that uses feature based description for the character image. First the matching regions are identified. Then based on some well-defined ratings of the assignments, the image elements are compared to the model. This procedure requires a search technique in a multidimensional space for finding the global maximum of some functions [1, 3]. A handwritten Chinese character system is proposed [1, 3] where a small number of critical structural features such as end points, hooks, T-shape, cross, and corner are used. The recognition is done by computing the matching probabilities between two features by a relaxation method. The matching techniques mentioned above are sometimes used individually or combined in many ways as part of the OCR schemes.

(b) Statistical techniques: The statistical decision theory is concerned with statistical decision functions and a set of optimality criteria which maximizes the probability of the observed pattern given the model of a certain class [1, 3]. The statistical techniques are mostly based on three major assumptions viz (i) The distribution of the feature set is Gaussian or in the worst case uniform (ii) There are sufficient statistics available for each class and (iii) Given ensemble of images $\{I\}$ one is able to extract a set of features $\{f_i\} \in F$; $i = \{1, \ldots, n\}$ which represents each distinct class of patterns. The measurements taken from n features of each character unit can be thought to represent an n-dimensional vector space and the vector whose coordinates correspond to the measurements taken represents the original character unit. The major statistical approaches applied in the CR field are: (i) nonparametric recognition (ii) parametric recognition (iii) clustering analysis (iv) hidden markov chains and (v) fuzzy set reasoning.

(i) Nonparametric recognition: This method is used to separate different pattern classes along hyperplanes defined in a given hyperspace. The best known method of nonparametric classification is ANN and is extensively used in OCR [1, 3]. It does not require apriori information about the data. An incoming pattern is classified using the cluster whose center is the minimum distance from the pattern over all the clusters.

(ii) Parametric recognition: Since apriori information is available about the characters in the training data, it is possible to obtain a parametric model for each character [1, 3]. Once the parameters of the model which are based on some probabilities are obtained, the characters are classified according to some decision rules such as maximum likelihood or bayes method.

(iii) Clustering analysis: The clusters of character features which represent distinct classes are analyzed by way of clustering methods. Clustering can be performed either by agglomerative or divisive algorithms. The agglomerative algorithms operate step-by-step merging of small clusters into larger ones by a distance criterion. On the other hand, the divisive methods split the character classes under certain rules for identifying the underlying character [1, 3].

(iv) Hidden markov chains: The hidden markov chains are the widely and successfully used technique for handwritten OCR problems [1, 3]. It is defined as a stochastic process generated by two interrelated mechanisms consisting of a markov chain having a finite number of states and a set of random functions each of which is associated with a state [1, 3]. At discrete instants of time the process is assumed to be in some state and an observation is generated by the random function corresponding to the current state. The underlying Markov chain then changes states according to its transitional probabilities. Here, the job is to build a model that explains and characterizes the occurrence of the observed symbols [1, 3]. The output corresponding to a single symbol can be characterized as discrete or continuous. The discrete outputs may be characters from a finite alphabet or quantized vectors from a codebook while continuous outputs are represented by samples from a continuous waveform. In generating a character, the system passes from one state to another each state emitting an output according to some probabilities until the entire character is obtained. There are two basic approaches to OCR systems using hidden markov chains such as model and path discriminant hidden markov chains [1, 3].

(v)   Fuzzy set reasoning: The fuzzy set reasoning employs fuzzy set elements in describing the similarities between the features of the characters. The fuzzy set elements give more realistic results when there is no a priori knowledge about the data and therefore the probabilities cannot be calculated. The characters can be viewed as a collection of strokes which are compared to reference patterns by fuzzy similarity measures. Since the strokes under consideration are fuzzy in nature, the fuzziness is utilized in the similarity measure. In order to recognize a character, an unknown input character is matched with all the reference characters and is assigned to the class of the reference character with the highest score of similarity among all the reference characters. The fuzzy similarity [1, 3] measure is utilized to define the fuzzy entropy for handwritten Chinese characters. A handwritten OCR system is proposed [1, 3] using a fuzzy graph theoretic approach where each character is described by a fuzzy graph. A fuzzy graph matching algorithm is then used for recognition. An algorithm is presented which uses average values of membership for final decision [1, 3].

(c) Structural techniques: The recursive description of a complex pattern in terms of simpler patterns based on the shape of the object was the initial idea behind the creation of the structural pattern recognition. These patterns are used to describe and classify the characters in OCR systems. The characters are represented as the union of the structural primitives. It is assumed that the character primitives extracted from writing are quantifiable and one can find the relations among them. The structural methods are applied to the OCR problems are: (i) grammatical methods and (ii) graphical methods.

(i)   Grammatical methods: The grammatical methods consider the rules of linguistics for analyzing the written characters. Later various orthographic, lexicographic and linguistic rules were applied to the recognition schemes. The grammatical methods create some production rules in order to form the characters from a set of primitives through formal grammars. These methods may combine any type of topological and statistical features under some syntactic and semantic rules [1, 3]. Formal tools like language theory allows to describe the admissible constructions and to extract the contextual information about the writing by using various types of grammars such as string grammars, graph grammars, stochastic grammars and picture description language [1, 3]. In grammatical methods, training is done by describing each character by a grammar G_i. In the recognition phase, the string, tree or graph of any character is analyzed in order to decide to which pattern grammar it belongs [1, 3]. The top down or bottom up parsing does syntax analysis. The grammatical methods in OCR area are applied in various character levels [1, 3]. In character level, picture description language is used to model each character in terms of a set of strokes and their relationship. This approach has been used for Indian OCR where Devanagari characters are presented by a picture description language [1, 3]. The system stores the structural description in terms of primitives and the relations. The recognition involves a search for the unknown character based on the stored description. The grammatical methods are mostly used in the post-processing stage for correcting the recognition errors [1, 3].

(ii) Graphical methods: The characters can also be represented by trees, graphs, digraphs or attributed graphs. The character primitives such as strokes are selected by a structural approach irrespective of how the final decision making is made in the recognition [1, 3]. For each class, a graph or tree is formed in the training stage to represent strokes. The recognition stage assigns the unknown graph to one of the classes by using a graph similarity measure. There are a variety of approaches that use the graphical methods. The hierarchical graph representation approach is used for handwritten Chinese OCR [1, 3]. Simon have proposed [1, 3] an off-line cursive script recognition scheme. The features are regularities which are defined as uninformative parts and singularities which are defined as informative strokes about the characters. The stroke trees are obtained after skeletonization. The goal is to match the trees of singularities. Although it is computationally expensive, relaxation matching is a popular method in graphical approaches to the OCR systems [1, 3].

(d) ANNs: The ANN possess a massively parallel architecture such that it performs computation at a higher rate compared to the classical techniques. It adapts to the changes in data and learns the characteristics of input signal. ANN contains many nodes. The output from one node is fed to another one in the network and the final decision depends on the complex interaction of all nodes. In spite of the different underlying principles, it can be shown that most of the ANN architectures are equivalent to statistical pattern recognition methods [1, 3]. Several approaches exist for training of ANNs [1, 3]. These include the error correction, boltzman, hebbian and competitive learning. They cover binary and continuous valued input as well as supervised and unsupervised learning. The ANN architectures are classified into two major groups viz feedforward and feedback (recurrent) networks. The most common ANNs used in the OCR systems are the multilayer perceptron of the feedforward networks and the kohonen's self-organizing map of the feedback networks. The multilayer perceptron proposed by Rosenblatt [1, 3] and elaborated by Minsky and Papert [1, 3] has been applied in OCR. An example is the feature recognition network proposed by Hussain and Kabuka [1, 3] which has a two-level detection scheme. The first level is for detection of sub-patterns and the second level is for detection of the characters. The neocognitron of Fukushima [1, 3] is a hierarchical network consisting of several layers of alternating neuron-like cells. S-cells are used for feature extracting and C-cells allow for positional errors in the features. The last layer is the recognition layer. Some of the connections are variable and are be modified by learning. Each layer of S and C cells are called cell planes. Here training patterns useful for deformation-invariant recognition of a large number of characters are selected. The feedforward ANN approach to the machine printed OCR problem has proven to be successful [1, 3] where ANN is trained with a database of 94 characters and tested in 300,000 characters generated by a postscript laser printer with 12 common fonts in varying size. Here Garland et al. propose a two-layer ANN trained by a centroid dithering process. The modular ANN architecture is used for unconstrained handwritten numeral recognition in [1, 3]. The whole classifier is composed of subnetworks. A subnetwork which contains three layers is responsible for a class among ten classes. Most of the recent developments on handwritten OCR research are concentrated on Kohonen's self-organizing map (SOM) [1, 3]. SOM integrates the feature extraction and recognition steps in a large training set of characters. It can be shown that it is analogous to k-means clustering algorithm. An example of SOM on OCR systems is the study by [1, 3] which presents a self-organization matching approach to accomplish the recognition of handwritten characters drawn with thick strokes. In [1, 3] a combination of modified SOM and learning vector quantization is proposed to define a three-dimensional ANN model for handwritten numeral recognition. Higher recognition rates are reported with shorter training time than other SOMs.

2.3.8 Post-processing

The eighth OCR component is post-processing. Some of the commonly used post-processing activities include grouping and error detection and correction. In grouping symbols in text are associated with strings. The result of plain symbol recognition in text is a set of individual symbols. However, these symbols do not usually contain enough information. These individual symbols are associated with each other making up words and numbers. The grouping of symbols into strings is based on symbols' location in document. The symbols which are sufficiently close are grouped together. For fonts with fixed pitch grouping process is easy as position of each character is known. For typeset characters distance between characters are variable. The distance between words are significantly large than distance between characters and grouping is therefore possible. The problems occur for handwritten characters when text is skewed. Until grouping each character is treated separately, the context in which each character appears has not been exploited. However, in advanced optical text recognition problems, system consisting only of single character recognition is not sufficient. Even best recognition systems will not give 100% correct identification of all characters [3, 4, 7]. Only some of these errors are detected or corrected by the use of context. There are two main approaches. The first utilizes the possibility of sequences of characters appearing together. This is done by using rules defining syntax of word. For different languages the probabilities of two or more characters appearing together in sequence can be computed and is utilized to detect errors. For example, in English language probability of k appearing after h in a word is zero and if such a combination is detected an error is assumed. Another approach is dictionaries usage which is most efficient error detection and correction method. Given a word in which an error is present and the word is looked up in dictionary. If the word is not in dictionary an error is detected and is corrected by changing word into most similar word. The probabilities obtained from classification helps to identify character erroneously classified. The error transforms word from one legal word to another and such errors are undetectable by this procedure. The disadvantage of dictionary methods is that searches and comparisons are time consuming.

2.4 Applications of Optical Character Recognition Systems

The last few decades have seen a widespread appearance of commercial OCR products satisfying requirements of different users. In this section we highlight some notable application areas of OCR. The major application areas are often distinguished as data entry, text entry and process automation. Interested readers can refer [3, 8, 11] for different OCR application areas.

The data entry area [7] covers technologies for entering large amounts of restricted data. Initially such machines were used for banking applications. The systems are characterized by reading only limited set of printed characters usually numerals and few special symbols. They are designed to read data like account numbers, customer's identification, article numbers, amounts of money etc. The paper formats are constrained with a limited number of fixed lines to read per document. Because of these restrictions, readers of this kind may have a very high throughput up to 150 documents per hour. Single character error and reject rates are 0.0001 and 0.01% respectively. Due to limited character set these readers are usually tolerant to bad printing quality. These systems are specially designed for their applications and prices are therefore high.

The text entry reading machines [7] are used as page readers in office automation. Here the restrictions on paper format and character set are exchanged for constraints concerning font and printing quality. The reading machines are used to enter large amounts of text, often in word processing environment. These page readers are in strong competition with direct key-input and electronic exchange of data. As character set read by these machines is rather large, the performance is extremely dependent on quality of the printing. However, under controlled conditions single character error and reject rates are about 0.01 and 0.1% respectively. The reading speed is few hundred characters per second.

In process automation [7] major concern is not to read what is printed but rather to control some particular process. This is actually automatic address reading technology for mail sorting. Hence, the goal is to direct each letter into appropriate bin regardless of whether each character was correctly recognized or not. The general approach is to read all information available and use postcode as redundancy check. The acceptance rate of these systems is obviously dependent on properties of mail. This rate therefore varies with percentage of handwritten mail. Although rejection rate for mail sorting may be large, miss rate is usually close to zero. The sorting speed is typically about 30 letters per hour.

The abovementioned application areas are those in which OCR has been successful and widely used. However, many other areas of applications exist and some of which are [3, 7]:

(a) Aid for blind: In the early days before digital computers and requirement for input of large amounts of data emerged this was an imagined application area for reading machines. Along with speech synthesis system such reader enables blind to understand printed documents.

(b) Automatic number plate readers: A few systems for automatic reading of number plates of cars exist. As opposed to other OCR applications, input image is not natural bilevel image and must be captured by very fast camera. This creates special problems and difficulties although character set is limited and syntax restricted.

(c) Automatic cartography: The character recognition from maps presents special problems within character recognition. The symbols are intermixed with graphics, text is printed at different angles and characters are of several fonts or even handwritten.

(d) Form readers: Such systems are able to read specially designed forms. In such forms all irrelevant information to reading machine is printed in colour invisible to scanning device. The fields and boxes indicating where to enter text is printed in this invisible colour. The characters are in printed or hand written upper case letters or numerals in specified boxes. The instructions are often printed on form as how to write each character or numeral. The processing speed is dependent on amount of data on each form but may be few hundred forms per minute. The recognition rates are seldom given for such systems.

(e) Signature verification and identification: This application is useful for banking environment. Such system establishes the identity of writer without attempting to read handwriting. The signature is simply considered as pattern which is matched with signatures stored in reference database.

2.5 Status of Optical Character Recognition Systems

A wide variety of OCR systems are currently commercially available [7]. In this section we explore the capabilities of OCR systems and the main problems encountered therein. We also take a step forward in discussing the evaluation performance of an OCR system.

OCR systems are generally divided into two classes [3]. The first class includes special purpose machines dedicated to specific recognition problems. The second class covers systems that are based on PC and low cost scanner. The first recognition machines are all hardwired devices. As these hardware were expensive, throughput rates were high to justify cost and parallelism was exploited. Today such systems are used in specific applications where speed is of high importance. For example, within areas of mail sorting and check reading. The cost of these machines are still high up to few million dollars and they recognize wide range of fonts. The advancements in computer technology has made it possible to fully implement recognition part of OCR in software packages which work on personal computers. The present PC systems are comparable to large scaled computers of early days and their cost of such systems are low. However, there are some limitations in such OCR software especially when it comes to speed and character sets read. The hand held scanners for reading do also exist. These are usually limited to reading of numbers and few additional letters or symbols of fixed fonts. They often read a line at a time and transmits it to application programs. A wide array of commercial software products are available over the years. The speed of these systems have grown over years. The sophistication of OCR system depends on type and number of fonts recognized [3]. Based on the OCR systems' capability to recognize different character sets, five different classes of systems are recognized viz. fixedfont, multifont, omnifont, constraint handwriting and scripts [7].

The fixedfont OCR machines [7] deal with recognition of one specific typewritten font. These fonts are characterized by fixed spacing between each character. In several standards fonts are specially designed for OCR, where each character has a unique shape to avoid ambiguity with other similar characters. Using these character sets it is quite common for commercial OCR machines to achieve recognition rate as high as 99.99% with high reading speed. The systems of first generation OCR were fixed font machines and the methods applied were based on template matching and correlation.

The multifont OCR machines [7] recognize more than one font. However, fonts recognized by these machines are usually of same type as those recognized by fixed font system. These machines appeared after fixedfont machines. They are able to read up to about ten fonts. The limit in number of fonts is due to pattern recognition algorithm and template matching which required that a library of bit map images of each character from each font was stored. The accuracy is quite good even on degraded images as long as fonts in library are selected with care.

An omnifont OCR machine [7] can recognize mostly non-stylized fonts without having to maintain huge databases of specific font information. Usually omnifont technology is characterized by feature extraction usage. The database of an omnifont system contains description of each symbol class instead of symbols themselves. This gives flexibility in automatic recognition of variety of fonts. Although omnifont is common for OCR systems, this should not be understood that system is able to recognize all existing fonts. No OCR machine performs equally well on all the fonts used by modern typesetters.

The recognition of constrained handwriting through OCR machine deals with the unconnected normal handwritten characters' problem. The optical readers with such capabilities are common these days [3, 4] and they exist. These systems require well-written characters and most of them recognize digits unless certain standards for hand printed characters are followed. The characters should be printed as large as possible to retain good resolution and are entered in specified boxes. The writer is instructed to keep to certain models avoiding gaps and extra loops. Commercially intelligent character recognition is often used for systems that recognize hand printed characters.

Generally all methods for character recognition described here deal with isolated character recognition problem. However, to humans it would be more interesting if it were possible to recognize entire words consisting of cursively joined characters. The script recognition deals with the problem of recognizing unconstrained handwritten characters which may be connected or cursive. In signature verification and identification the objective is to establish identity of writer irrespective of handwritten contents. The identification establishes identity of writer by comparing specific attributes of pattern describing the signature with list of writers stored in a reference database. When performing signature verification the claimed identity of writer is known and the signature pattern is matched against the signature stored in database for the person. Many such systems of this type are commercially available [4]. A more challenging problem is script recognition where contents of handwriting must be recognized. The variations in shape of handwritten characters are infinite and depend on writing habit, style, education, mood, social environment and other conditions of writer. Even best trained optical readers and humans make about 4% errors when reading. The recognition of characters written without any constraint is available in some commercially available systems [3].

The accuracy of OCR systems directly depends upon the quality of input documents [2, 7]. The major problems encountered in different documents are classified in terms of (a) shape variations due to serifs and style variations (b) deformations caused by broken characters, smudged characters and speckle (c) spacing variations due to subscripts, superscripts, skew and variable spacing and (d) a mixture of text and graphics. These imperfections affect and cause problems in different parts of the recognition process of OCR systems resulting in rejections or misclassifications.

The majority of errors in OCR systems are often due to problems in scanning process and the following segmentation which results in joined or broken characters [3]. The segmentation process errors also results in confusion between text and graphics or between text and noise. Even if a character is printed, scanned and segmented correctly it may be incorrectly classified. This happens if character shapes are similar and selected features are not sufficiently efficient in separating different classes or if the features are difficult to extract and has been computed incorrectly. The incorrect classification also happens due to poor classifier design. This occurs if the classifier has not been trained on sufficient test samples representing all the possible forms of each character. The errors may also creep in due to post processing when isolated symbols are associated to reconstruct the original words as characters which are incorrectly grouped. These problems occur if the text is skewed such that in some cases there is proportional spacing and symbols have subscripts or superscripts. As OCR devices employ wide range of approaches to character recognition all systems are not equally affected by the abovementioned complexities [7]. The different systems have their strengths and weaknesses. In general, however the problems of correct segmentation of isolated characters are the ones most difficult to overcome and recognition of joined and split characters are usually weakest link of any OCR system.

Finally we conclude this section with some insights to the performance evaluation of OCR systems [3]. There exist no standardized test sets for character recognition. This is mainly because the performance of OCR system is highly dependent on the quality of input which makes it difficult to evaluate and compare different systems. The recognition rates are often given and usually presented as percentage of characters correctly classified. However, this does not say anything about the errors committed. Thus in evaluation of OCR system three different performance rates are investigated such as (a) recognition rate which is the proportion of correctly classified characters (b) rejection rate which is the proportion of characters which the system is unable to recognize and (c) error rate which is the proportion of characters erroneously classified. There is usually a trade-off between different recognition rates. A low error rate leads to higher rejection rate and a lower recognition rate. Because of the time required to detect and correct OCR errors, error rate is most important when evaluating the cost effectiveness of an OCR system. The rejection rate is less critical. As an example consider reading operation from barcode. Here a rejection while reading bar-coded price tag will only lead to rescanning of the code whereas a wrongly decoded price tag results in charging wrong amount to the customer. In barcode industry error rates are therefore as low as one in million labels whereas rejection rate of one in a hundred is acceptable. In view of this it is apparent that it is not sufficient to look solely on recognition rates of a system. A correct recognition rate of 99% implies an error rate of 1%. In case of text recognition on printed page which on average contains about 2000 characters, an error rate of 1% means 20 undetected errors per page. In postal applications for mail sorting where an address contains about 50 characters, an error rate of 1% implies an error on every other piece of mail.

2.6 Future of Optical Character Recognition Systems

All through the years, the methods of OCR systems have improved from primitive schemes suitable only for reading stylized printed numerals to more complex and sophisticated techniques for the recognition of a great variety of typeset fonts [4] and also hand printed characters. The new methods for character recognition continue appear with development of computer technology and decrease in computational restrictions [3]. However, the greatest potential lies in exploiting existing methods by hybridizing technologies and making more use of context. The integration of segmentation and contextual analysis improves recognition of joined and split characters. Also higher level contextual analysis which looks at semantics of entire sentences are useful. Generally there is a potential in using context to larger extent than what is done today. In addition, a combination of multiple independent feature sets and classifiers where weakness of one method is compensated by the strength of another improves recognition of individual characters [2]. The research frontiers within character recognition continue to move towards recognition of sophisticated cursive script that is handwritten connected or calligraphic characters. Some promising techniques within this area deal with recognition of entire words instead of individual characters.

References

1. Arica, N., Vural, F. T. Y., An Overview of Character Recognition focused on Offline Handwriting, IEEE Transactions on Systems, Man and Cybernetics – Part C: Applications and Reviews, 31(2), pp 216–233, 2001.
2. Bunke, H., Wang, P. S. P. (Editors), Handbook of Character Recognition and Document Image Analysis, World Scientific, 1997.
3. Chaudhuri, A., Some Experiments on Optical Character Recognition Systems for different Languages using Soft Computing Techniques, Technical Report, Birla Institute of Technology Mesra, Patna Campus, India, 2010.
4. Cheriet, M., Kharma, N., Liu, C. L., Suen, C. Y., Character Recognition Systems: A Guide for Students and Practitioners, John Wiley and Sons, 2007.
5. Dholakia, K., A Survey on Handwritten Character Recognition Techniques for various Indian Languages, International Journal of Computer Applications, 115(1), pp 17–21, 2015.
6. Mantas, J., An Overview of Character Recognition Methodologies, Pattern Recognition, 19(6), pp 425–430, 1986.
7. Rice, S. V., Nagy, G., Nartker, T. A., Optical Character Recognition: An Illustrated Guide to the Frontier, The Springer International Series in Engineering and Computer Science, Springer US, 1999.

8. Schantz, H. F., The History of OCR, Recognition Technology Users Association, Manchester Centre, VT, 1982.
9. Scurmann, J., Reading Machines, Proceedings of International Joint Conference on Pattern Recognition, Munich, pp 1031–1044, 1982.
10. Singh, S., Optical Character Recognition Techniques: A Survey, Journal of Emerging Trends in Computing and Information Sciences, 6 (4), pp 545–550, 2013.
11. Young, T. Y., Fu, K. S., Handbook of Pattern Recognition and Image Processing, Academic Press, 1986.
12. Yu, F. T. S., Jutamulia, S. (Editors), Optical Pattern Recognition, Cambridge University Press, 1998.

Chapter 3
Soft Computing Techniques for Optical Character Recognition Systems

Abstract The continuous increase in demand to discover robust and low cost optical character recognition (OCR) systems has prompted researchers to look for rigorous methods of character recognition. In the past OCR systems have been built through traditional pattern recognition and machine learning approaches. There has always been a quest to develop best OCR products which satisfy the user's needs. Since past few decades soft computing techniques have come up as a promising candidate for the development of cost effective OCR systems. Some important soft computing techniques for optical character recognition (OCR) systems are presented in this chapter. They are hough transform for fuzzy feature extraction, genetic algorithms (GA) for feature selection, fuzzy multilayer perceptron (FMLP), rough fuzzy multilayer perceptron (RFMLP), fuzzy support vector machine (FSVM), fuzzy rough versions of support vector machine (FRSVM), hierarchical fuzzy bidirectional recurrent neural networks (HFBRNN) and fuzzy markov random fields (FMRF). These techniques are used for developing OCR systems for different languages viz English, French, German, Latin, Hindi and Gujrati languages. The soft computing methods are used in the different steps of OCR systems discussed in Chap. 2. A comprehensive assessment of these methods is performed in Chaps. 4–9 for the stated languages. A thorough understanding of this chapter will help the readers to appreciate the reading material presented in the abovementioned chapters.

Keywords Soft computing · OCR · FMLP · RFMLP · FSVM · FRSVM · HFBRNN · FMRF

3.1 Introduction

The ever increasing demand to discover robust and low cost optical character recognition (OCR) systems to withstand tough competition has prompted researchers to look for rigorous methods of character recognition [5, 10, 22]. In the past OCR

© Springer International Publishing AG 2017
A. Chaudhuri et al., *Optical Character Recognition Systems for Different Languages with Soft Computing*, Studies in Fuzziness and Soft Computing 352,
DOI 10.1007/978-3-319-50252-6_3

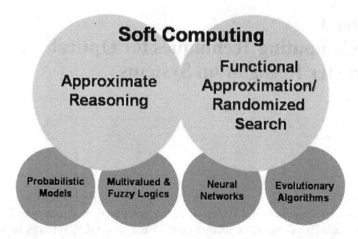

Fig. 3.1 The computational constituents in soft computing

systems have been built through traditional pattern recognition and machine learning approaches [5]. However, in order to increase the marketability of OCR systems in terms of economy and efficiency, there has always been a quest to design and produce OCR products which satisfy the user's need and requirements. As search for the best OCR systems has always fascinated mankind, several strategies and methods have been attempted and devised for searching the best possible solution. In this direction soft computing [14, 20] has emerged as a promising candidate since the past few decades.

Soft computing is an important and challenging mathematical aspect in pattern recognition and artificial intelligence domain. It is different from the conventional hard computing that normally happens in terms of, approximation or exact solution that is to be obtained for any given problem. It is generally used for NP complete problems [9] in generic as no specific polynomial time algorithms can be obtained for them. Soft computing is often compared with the understanding involved in the human mind.

Soft computing has been applied to almost all disciplines of engineering and science [9] with appreciable success. It refers to a collection of computational techniques from machine learning and artificial intelligence which attempt to study, model and analyze very complex phenomena [14, 20] for which conventional methods have not yielded low cost, analytic and complete solutions. Thus, the major goal of soft computing is to develop intelligent machines to provide solutions to the real world problems which are difficult to be modeled mathematically [9]. It involves in finding the best solution in most effective way to given problem eventually with certain constraints. The Fig. 3.1 gives a schematic structure of the important computational constituents involved in soft computing framework [9, 14]. The principal constituents of soft computing are machine learning, fuzzy logic, probabilistic reasoning, neural computing or artificial neural networks (ANN) and evolutionary computing or evolutionary algorithms (EA). For interested readers further details are available in [14].

Soft computing has some uniqueness with the results are derived from experimental data, generating approximate outputs for unknown inputs using previously processed outputs from known inputs. The entire computational process in the soft computing framework can be described as a transformation from modeling space M to solution space S and finally to decision space D [9]:

$$M \to S \to D$$

Here, the mapping function $\oslash : S \to D$ is decision function and elements $d \in D$ are decisions. The earlier approaches could model and precisely analyze only relatively simple systems. More complex systems arising in biology, medicine, humanities, management science etc. often remained intractable to conventional mathematical and analytical methods. It is realized that complex real world problems require intelligent systems that combine knowledge, techniques and methodologies from various sources. These intelligent systems are supposed to possess human like expertise within a specific domain, adapt themselves and learn to do better in changing environments and explain how they make decisions or take actions [9]. In real world computing problems, it is advantageous to use several computing techniques synergistically rather than exclusively, resulting in construction of hybrid intelligent systems such as soft computing.

The soft computing approach thus parallels the remarkable ability of human mind to reason and learn in an environment of uncertainty and imprecision [9, 14]. Soft computing differs from conventional or hard computing in that unlike hard computing it is tolerant of imprecision, uncertainty, partial truth and approximation. The role model for soft computing is human mind. The guiding principle of soft computing is [9, 14]:

Exploit tolerance for imprecision, uncertainty, partial truth and approximation to achieve tractability, robustness and low cost solution

The basic ideas underlying soft computing in its current incarnation have links to many earlier influences. Among them Zadeh's 1965 paper on fuzzy sets [25], 1973 paper on analysis of complex systems and decision processes [9] and 1979 report (1981 paper) on possibility theory and soft data analysis [24] are worth mentioning. The inclusion of ANN and genetic algorithms (GA) [12] in soft computing came at later point. As represented in Fig. 3.1, the principal constituents of soft computing [14, 20] are thus fuzzy sets, ANN, EA, machine learning and probabilistic reasoning with latter subsuming belief networks, chaos theory and parts of learning theory [9]. Each of these constituents has its own strength. What is important to note is that soft computing is not a mélange. Rather, it is partnership in which each of the partners contributes distinct methodology for addressing problems in its domain. In this perspective, principal constituent methodologies in soft computing are complementary rather than competitive. Furthermore, soft computing may be viewed as foundation component for computational intelligence. Some exciting and important hybrid soft computing tools available since past few decades are rough-fuzzy, neuro-fuzzy, neuro-fuzzy-genetic, rough-fuzzy-genetic etc. [9, 14]. Keeping in view the computational complexity involved in the development of OCR systems in real life situations, the focus is to design optimal algorithms using different soft computing concepts. MATLAB is used to model the various implementations. The developed methods and algorithms have appreciable computational complexity. The new models adapt themselves to new situations dynamically which makes it possible to apply them to other complex situations.

In this chapter we present some important soft computing techniques like Hough transform for fuzzy feature extraction, GA for feature selection, rough fuzzy multilayer perceptron (RFMLP) [7, 8, 17] fuzzy and fuzzy rough versions of support vector machines (SVM) [2–5] hierarchical fuzzy bidirectional recurrent neural networks (HFBRNN) [1, 6] and fuzzy markov random fields (FMRF) [6, 26] towards developing OCR systems for different languages viz english, french, german, latin, hindi and gujrati languages. These methods are used in the different steps of OCR systems discussed in Chap. 2. A comprehensive assessment of these methods is performed in Chaps. 4–9 for the stated languages. A thorough understanding of this chapter will help the readers to appreciate the reading material presented in abovementioned chapters.

This chapter is organized as follows. In Sect. 3.2 some important constituents of soft computing are presented. In Sect. 3.3 Hough transform for fuzzy feature extraction is discussed. This is followed by a discussion of GA for feature selection in Sect. 3.4. The next section presents RFMLP. In Sect. 3.6 the fuzzy and fuzzy rough versions of SVM are presented. The HFBRNN are discussed in Sect. 3.7. The next section highlights the FMRF. Finally, the chapter concludes with some important directions to other soft computing techniques.

3.2 Soft Computing Constituents

In this section we present some important constituents of soft computing used for OCR systems viz fuzzy sets, ANN, GA and rough sets. This will help the readers to appreciate the several soft computing techniques presented in the rest of the chapter. Interested readers can refer [9, 14, 23] for further insights and details into these soft computing techniques.

3.2.1 Fuzzy Sets

The human brain interprets imprecise and incomplete sensory information provided by perceptive organs. Fuzzy set theory [25] provides systematic calculus to deal with such information linguistically and it performs numerical computation by using linguistic labels stipulated by membership functions. The logic revolving fuzzy sets deals with approximate reasoning. Here, the degrees of truth replace probabilities. The membership function denoting linguistic labels are fuzzy truth values which represent membership in vaguely defined sets. The set membership

Fig. 3.2 The variable temperature represented through various degrees of fuzziness

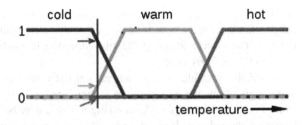

values lie in the range between 0 (false) and 1 (true) inclusively. They represent the linguistic form of imprecise concepts like slightly, quite, very etc. Specifically it allows partial membership in set. Fuzzy logic is the generalized version of the classical logic which only permits conclusions which are either true (1) or false (0). In real life often there are propositions with variable answers. For example, one might find when asking a group of people to identify a color where the truth appears as the result of reasoning from inexact or partial knowledge in which the sampled answers are mapped on a spectrum.

The humans often operate using fuzzy evaluations in many everyday situations. In the case where someone is tossing an object into a container from a distance, the person does not compute exact values for the object weight, density, distance, direction, container height and width and air resistance to determine the force and angle to toss the object. Instead the person instinctively applies quick fuzzy or vague estimates based upon previous experience to determine what output values of force, direction and vertical angle to use to make the toss.

Both the degrees of truth and probabilities range between 0 and 1 and hence may seem similar at first [21]. For example, let a 100 ml glass contain 30 ml of water. Then we may consider two concepts viz empty and full. The meaning of each of them can be represented by a certain fuzzy set. Then one might define the glass as being 0.7 empty and 0.3 full. Note that the concept of emptiness would be subjective and thus would depend on the observer or designer. Another designer might, equally well design a set membership function where the glass would be considered full for all values down to 50 ml. It is essential to realize that fuzzy logic uses degrees of truth as a mathematical model of vagueness, while probability is a mathematical model of ignorance.

Any basic application might characterize various sub ranges of a continuous variable. For instance, a temperature measurement for antilock brakes might have several separate membership functions defining particular temperature ranges needed to control the brakes properly. Each function maps the same temperature value to a truth value in the 0–1 range. These truth values can then be used to determine how the brakes should be controlled. The Fig. 3.2 represents the variable temperature through various degrees of fuzziness. In Fig. 3.2 the meanings of the expressions *cold*, *warm*, and *hot* are represented by functions mapping a temperature scale. A point on that scale has three truth values which represent the various degrees of fuzziness: one for each of the three functions. The vertical line in the image represents a particular temperature that the three arrows (truth values) gauge. Since the red arrow points to zero, this temperature may be interpreted as *not hot*. The orange arrow at 0.2 may describe it as *slightly warm* and the blue arrow at 0.8 as *fairly cold*.

While the variables in mathematics usually take numerical values; in fuzzy logic applications the non-numeric aspects are often used to facilitate the expression of rules and facts [15]. A linguistic variable may be *young* or its antonym *old*. However, the value of linguistic variables is that they can be modified via linguistic hedges applied to primary terms. These linguistic hedges can be associated with certain functions. The fuzzification operations map mathematical input values into fuzzy membership functions whereas the defuzzificaion operations map a fuzzy output membership functions into a crisp output value [27] that is used for decision or control purposes.

Fuzzy sets thus move forward to generalize classical two-valued logic for reasoning under uncertainty. This is achieved when notation of membership in set becomes a matter of degree. By doing this, two things are accomplished viz ease of describing human knowledge involving vague concepts and enhanced ability to develop cost-effective solution to real-world problems. Fuzzy sets are based on multi-valued logic which is model-less approach and a clever disguise of probability theory [24, 25]. Fuzzy sets provide an effective means of describing behavior of systems which are too complex or too ill-defined to admit precise mathematical analysis by classical methods. They have shown enormous promise in handling uncertainties to a reasonable extent, particularly in decision making models under different kinds of risks, subjective judgment, vagueness and ambiguity.

3.2.2 Artificial Neural Networks

ANN [13] mimics the human nervous system especially the human brain. ANN are similar to biological neural networks in the performing by its units of functions collectively and in parallel, rather than by a clear delineation of subtasks to which individual units are assigned. In machine learning ANNs are a family of models inspired by biological neural networks which are used to estimate or approximate functions that can depend on a large number of inputs and are generally unknown. ANNs are represented through nodes that are connected by links. Each node performs simple operation to compute its output from its input which is transmitted through links connected to other nodes. The nodes correspond to neurons in brain and links correspond to synapses that transmit signal between neurons. Each link has a strength that is expressed by weight value. One of the major features of ANN is its learning capability from real life patterns or examples. They are typically specified using three fundamental features:

(a) Architecture that specifies what variables are involved in the network and their topological relationships. For example, the variables involved in

ANN might be the weights of the connections between the neurons along with activities of the neurons.

(b) Activity rule in ANN models refer to the time scale dynamics. Most ANN have short time scale dynamics. The local rules define how the activities of the neurons change in response to each other. Typically the activity rule depends on the weights or the parameters in the network.

(c) Learning rule specifies the way in which the neural network's weights change with time. This learning is usually viewed as taking place on a longer time scale than the time scale of the dynamics under the activity rule. Usually the learning rule will depend on the activities of the neurons. It may also depend on the values of the target values supplied by a teacher and on the current value of the weights.

For example, an ANN for handwriting recognition is defined by a set of input neurons which may be activated by the pixels of an input image. After being weighted and transformed by a function which is determined by the network's designer, the activations of these neurons are then passed on to other neurons. This process is repeated until finally, the output neuron that determines which character was read is activated. Like other machine learning methods, ANN have been used to solve a wide variety of tasks, like computer vision and speech recognition that are hard to solve using ordinary rule based programming.

While the details of learning algorithms of ANN vary from architecture to architecture, they have one common aspect viz they can adjust parameters in ANN so that the network learns to improve its performance. The most common forms of learning used in ANN are supervised and unsupervised learning [13]. The supervised learning is guided by specifying for each training input pattern, the class to which pattern is supposed to belong. Thus the desired response of network is used in learning algorithm for appropriate adjustment of weights. These adjustments are made incrementally in desired direction to minimize difference between desired and actual outputs which facilitates solution convergence [9, 13]. In unsupervised learning, network from its own classification of patterns. The classification is based on commonalities in certain features of input patterns. These require that network implementing an unsupervised learning be able to identify common features across range of input patterns. The major advantage of ANN is their flexible non-linear modeling capability and no need to specify particular model form [9]. Rather the model is adaptively formed based on the features presented in data. However, ANN require large amount of data in order to yield accurate results. No definite rule exists for the sample size requirement of given problem. The amount of data for network training depends on network structure, training method and complexity of particular problem or amount of noise in data on hand. With large enough sample, ANN can model any complex structure in data. ANN Some commonly used categories of ANN are single layer perceptron network, feed-forward network, radial basis function network, multi-layer perceptron network, kohonen self-organizing network, recurrent network, stochastic network, associative networks etc. [13]. The Fig. 3.3 shows multi-layer ANN that is interconnected through a group of nodes.

Fig. 3.3 A multi-layer ANN
interconnected through a
group of nodes

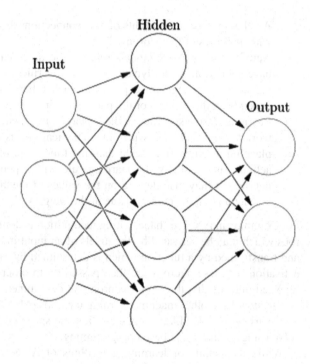

The modern software implementations of artificial neural networks, the approach inspired by biology has been largely abandoned for a more practical approach based on statistics and signal processing. In some of these systems, ANN or parts of ANN like artificial neurons form components in larger systems that combine both adaptive and non-adaptive elements. While the more general approach of such systems is more suitable for real-world problem solving, it has little to do with the traditional, artificial intelligence connectionist models. What they do have in common, however, is the principle of non-linear, distributed, parallel and local processing and adaptation. Historically, the use of neural network models marked a directional shift in the late eighties from high-level symbolic artificial intelligence, characterized by expert systems with knowledge embodied in if-then rules, to low-level sub-symbolic machine learning, characterized by knowledge embodied in the parameters of a dynamical system.

3.2.3 Genetic Algorithms

In artificial intelligence, GA is a search heuristic that mimics the process of natural selection. GA was first suggested by John Holland. This heuristic also sometimes called a metaheuristic is routinely used to generate useful solutions to optimization and search problems. GA belong to the larger class of EA which generate solutions to optimization problems using techniques inspired by natural evolution such as inheritance, mutation, selection and crossover. GA are unorthodox optimization search algorithms. They are nature inspired algorithms mimicking natural evolution [12].

In GA a population of candidate solutions is often represented through various abstractions called individuals, creatures or phenotypes to the optimization problem. These solutions evolve the optimization problem towards better solutions. Each candidate solution has a set of properties or chromosomes which can be mutated and altered. Traditionally, the solutions are represented in binary as strings of 0s and 1 s but other encodings are also possible. GA perform directed random search through a given set of alternatives with the aim of finding best alternative. Traditionally, solutions are represented in binary as strings of 0 and 1.

The evolution usually starts from a population of randomly generated individuals and is an iterative process with the population in each iteration called a generation. In each generation, the fitness of every individual in the population is evaluated. The fitness is usually the value of the objective function in the optimization problem being solved. The more fit individuals are stochastically selected from the current population and each individual's genome is modified which may sometimes be recombined and possibly randomly mutated to form a new generation. The new generation of candidate solutions is then used in the next iteration of the algorithm. Commonly, the algorithm terminates when either a maximum number of generations has been produced or a satisfactory fitness level has been reached for the population. A typical GA requires genetic representation of the following aspects [16]:

(a)   A genetic representation of the solution domain
(b)   A fitness function to evaluate the solution domain.

The solution is represented through an array of bits. These genetic representations have their parts are easily aligned due to their fixed size that facilitates simple crossover operation. The variable length representations may also be used, but crossover implementation is more complex in this case. Tree-like representations are explored in genetic programming and graph-form representations are explored in evolutionary programming; a mix of both linear chromosomes and trees is explored in gene expression programming. The fitness function is defined over genetic representation and measures quality of represented solution. Once genetic representation and fitness function are defined GA proceeds to initialize a population of solutions randomly and improve it through repetitive application of mutation, crossover, inversion and selection operators. The Figs. 3.4 and 3.5 represent a typical evolution flow [12] and the generation wise evolution in GA [12] respectively.

During each successive generation, a proportion of the existing population is selected to breed a new generation. Individual solutions are selected through a fitness based process where fitter solutions as measured by a fitness function are typically more likely to be selected. Certain selection methods rate the fitness of each solution and preferentially select the best solutions. Other methods rate only a random sample of the population as the former process may be very time consuming.

Fig. 3.4 The genetic
algorithm evolution flow

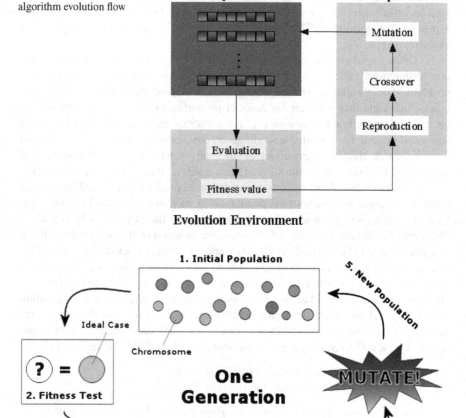

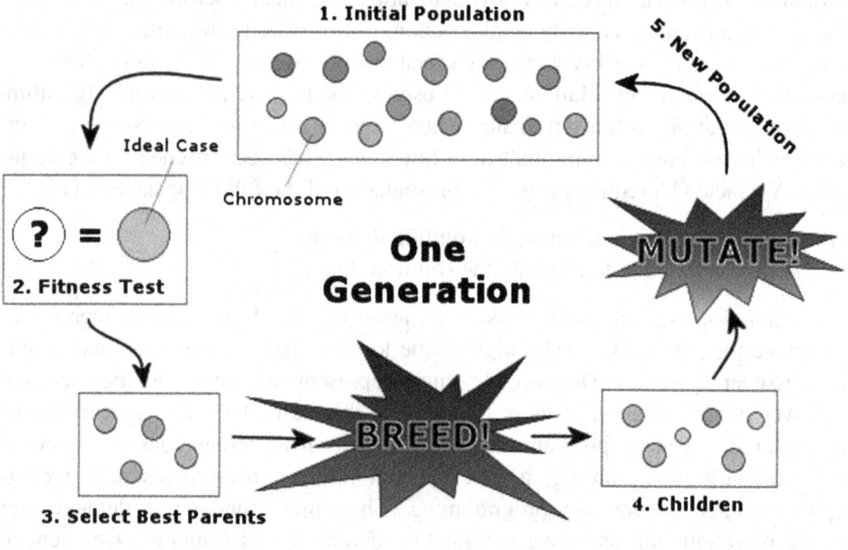

Fig. 3.5 The generation wise evolution in genetic algorithm

The fitness function is defined over the genetic representation and measures
the quality of the represented solution. The fitness function is always problem
dependent. For instance, in the knapsack problem one wants to maximize the total
value of objects that can be put in a knapsack of some fixed capacity. A representation of a solution might be an array of bits, where each bit represents a different
object and the value of the bit (0 or 1) represents whether or not the object is in the
knapsack. Not every such representation is valid, as the size of objects may exceed
the capacity of the knapsack. The fitness of the solution is the sum of values of all

objects in the knapsack if the representation is valid or 0 otherwise. In some problems, it is hard or even impossible to define the fitness expression; in these cases, a simulation may be used to determine the fitness function value of a phenotype. For example, computational fluid dynamics is used to determine the air resistance of a vehicle whose shape is encoded as the phenotype or even interactive genetic algorithms are used. The generational process in GA is repeated until a termination condition has been reached. Common terminating conditions are [12]:

(a) A solution is found that satisfies minimum criteria
(b) A fixed number of generations reached
(c) The allocated budget computation time or money is exhausted
(d) The highest ranking solution's fitness is reaching or has reached a plateau such that successive iterations no longer produce better results
(e) Through some manual inspection
(f) Some combinations of the above.

GA are simple to implement but their behavior is difficult to understand. In particular it is difficult to understand why these algorithms frequently succeed at generating solutions of high fitness when applied to practical problems. The building block hypothesis consists of the following aspects [16]:

(a) A description of a heuristic that performs adaptation by identifying and recombining building blocks i.e. low order, low defining length schemata with above average fitness.
(b) A hypothesis that a genetic algorithm performs adaptation by implicitly and efficiently implementing this heuristic.

The heuristic is often described as follows [16]:

(a) Short, low order and highly fit schemata are sampled, recombined or crossed over and resampled to form strings of potentially higher fitness. In a way by working with these particular schemata the complexity of the problem is reduced. Instead of building high performance strings by trying every conceivable combination, better and better strings are constructed from the best partial solutions of past samplings.
(b) Because highly fit schemata of low defining length and low order play such an important role in the action of genetic algorithms, they have been renamed as the building blocks. Just as a child creates magnificent fortresses through the arrangement of simple blocks of wood, so does a genetic algorithm seek near optimal performance through the juxtaposition of short, low-order, high-performance schemata, or building blocks.

3.2.4 Rough Sets

Rough sets were introduced by Pawlak [19]. They have emerged as another major mathematical approach for managing uncertainty arising from inexact, noisy or incomplete information [19]. Rough sets are methodologically significant in cognitive sciences especially in representation and reasoning with vague and imprecise knowledge, data classification, data analysis, learning and knowledge discovery [9]. The rough set theory has proved to be of substantial importance in many applications areas. The major distinction between rough and fuzzy sets is that former requires no external parameters and uses only information present in given data [9]. The fuzzy set theory hinges on notion of membership function on domain of discourse, assigning to each object grade of belongingness in order to represent an imprecise concept. The focus of rough set theory is on ambiguity caused by limited discernibility of objects in domain of discourse. The idea is to approximate any concept by pair of exact sets called lower and upper approximations. On basis of lower and upper approximations of rough set, the accuracy of approximating rough set can be calculated as ratio of cardinality of lower and upper approximations [19].

The rough set revolves around the concept of an information system which is defined by a pair $S = \langle U, A \rangle$, where U is a nonempty finite set called the universe and A is a nonempty finite set of attributes. An attribute a can be regarded as a function from the domain U to some value set V_a. A decision system is any information system of the form $A = (U, A \cup \{d\})$, where $d \notin A$ is the decision attribute. The elements of A are called conditional attributes. An information system can be represented as an attribute-value table, in which rows are labeled by objects of the universe and columns by the attributes. Similarly, a decision system can be represented by a decision table. With every subset of attributes $B \subseteq A$, an equivalence relation I_B can easily be associated on U, $I_B = \{(x, y) \in U: \forall a \in B, a(x) = a(y)\}$. Then, $I_B = \cap_{a \in B} I_a$. If $X \subseteq U$, the sets $\{x \in U: [x]_B \subseteq X\}$ and $\{x \in U: [x]_B \cap X = \emptyset\}$, where, $[x]_B$ denotes the equivalence class of the object $x \in U$ relative to I_B, are called the B-lower and B-upper approximation of X in S and denoted by $\underline{B}X, \overline{B}X$ respectively. $X \subseteq U$ is B-exact or B-definable in S if $\underline{B}X = \overline{B}X$. It may be observed that $\underline{B}X$ is the greatest B-definable set contained in X and $\overline{B}X$ is the smallest B-definable set containing X.

Another relevant aspect in rough sets is related to the knowledge reduction. The aim is to obtain irreducible but essential parts of the knowledge encoded by the given *information system*, which constitutes the *reducts* of the system. This in effect reduces to looking for *maximal* sets of attributes taken from the initial set A which induce the same partition on the domain as A. In other words, the essence of the information remains intact and superfluous attributes are removed. Reducts have already been characterized by the *discernibility matrices* and *discernibility functions*. Consider $U = \{x_1,...,x_n\}$ and $A = \{a_1,...,a_m\}$ in the *information system* $S = < U, A >$. By the *discernibility matrix* M(S), of S is meant an $n \times n$ matrix such that:

$$c_{ij} = \{a \in A : a(x_i) \neq a(x_j)\} \tag{3.1}$$

A discernibility function f_S is a function of m Boolean variables $\bar{a}_1, \ldots, \bar{a}_m$ corresponding to the attributes $a_1,...,a_m$ respectively and is defined as follows:

$$f_S(\bar{a}_1, \ldots, \bar{a}_m) = \Lambda\{\vee(c_{ij}) : 1 \leq i,j \leq n, j < i, c_{ij} \neq \Phi\} \tag{3.2}$$

In Eq. (3.2), $\vee(c_{ij})$ is the disjunction of all variables $\bar{a}$ with $a \in c_{ij}$. It is observed that $\{a_{i_1}, \ldots, a_{i_p}\}$ is a reduct in S if and only if $a_{i_1} \wedge \cdots \wedge a_{i_p}$ is a prime implicant of f_S.

In granular universe there may well be aspects which are conceptually imprecise in the sense that they are not represented by crisp subsets. This led to a direction where both rough sets and fuzzy sets can be integrated. The aim is to develop model of uncertainty stronger than either. Research work combining fuzzy and rough sets for developing efficient methodologies and algorithms towards various real life decision making applications have already appeared [9]. The integration of these theories with ANN is performed with the aim of building more efficient intelligent systems in soft computing paradigm.

3.3 Hough Transform for Fuzzy Feature Extraction

In this section Hough transform for fuzzy feature extraction in OCR systems is presented. Hough transform is a method for detection of lines and curves from images [6]. The basic Hough transform is generalized through fuzzy probabilistic concepts [24, 25]. Here, fuzzy Hough transform is presented where image points are treated as fuzzy points. The Hough transform for line detection uses mapping $r = x \cos\theta + y \sin\theta$ which provides three important characteristics of line in an image pattern. The parameters r and θ specify position and orientation of line. The count of (r, θ) accumulator cell used in Hough transform implementation specifies number of black pixels lying on it. With this in mind, we define a number of fuzzy sets on (r, θ) accumulator cells. Some notable fuzzy set definitions are shown in Table 3.1 for θ values in first quadrant. The definitions are extended for other values of θ. The fuzzy sets viz. long_line and short_line extract length information of different lines in pattern. The nearly_horizontal, nearly_vertical and slant_line represent skew and near_top, near_bottom, near_vertical_centre, near_right, near_left and near_horizontal_centre extract position information of these lines. The characteristics of different lines in an image pattern are mapped into properties of these fuzzy sets. For interested readers further details are available in [6].

Based on basic fuzzy sets [25], fuzzy sets are further synthesized to represent each line in pattern as combination of its length, position and orientation using t-norms [27]. The synthesized fuzzy sets are defined as long_slant_line $\equiv$ t-norm (slant_line, long_line), short_slant_line $\equiv$ t-norm (slant_line, short_line), nearly_vertical_long_line_near_left $\equiv$ t-norm (nearly_vertical_line, long_line, near_left_border). Similar basic fuzzy sets such as large_circle, dense_circle, centre_near_top etc. and synthesized fuzzy sets such as small_dense_circle_near_top, large_dense_circle_near_centre etc. are defined on (p, q, t) accumulator cells for circle extraction using Hough transform $t = \sqrt{(x-p)^2 + (y-q)^2}$. For a circle extraction (p, q) denotes origin, c is radius and count specifies the number of pixels lying on circle. A number of t-norms are available as fuzzy intersections among which standard intersection t-norm $(p, q) \equiv \min(p, q)$. For other pattern recognition

Table 3.1 Fuzzy set membership functions defined on Hough transform accumulator cells for line detection (x and y denote height and width of each character pattern)

Fuzzy set	Membership function
long_line	$\frac{cellcount}{\sqrt{x^2+y^2}}$
short_line	$2(\text{long\_line})$ if $count \leq \sqrt{x^2+y^2}/2$ $2(1 - \text{long\_line})$ if $count > \sqrt{x^2+y^2}/2$
nearly_horizontal_line	$\frac{\theta}{90}$
nearly_vertical_line	$1 - \text{nearly\_horizontal\_line}$
slant_line	$2(\text{nearly\_horizontal\_line})$ if $\theta \leq 45$ $2(1 - \text{nearly\_horizontal\_line})$ if $\theta > 45$
near_top	r/x if nearly_horizontal_line > nearly_vertical_line 0 otherwise
near_bottom	$(1 - \text{near\_top})$ if nearly_horizontal_line > nearly_vertical_line 0 otherwise
near_vertical_centre	$2(\text{near\_top})$ if ((nearly_horizontal_line > nearly_vertical_line) and $(r \leq x/2)$) $2(1 - \text{near\_top})$ if ((nearly_horizontal_line > nearly_vertical_line) and $(r > x/2)$) 0 otherwise
near_right_border	r/y if nearly_vertical_line > nearly_horizontal_line 0 otherwise
near_left_border	$(1 - \text{near\_right\_border})$ if nearly_vertical_line > nearly_horizontal_line 0 otherwise
near_horizontal_centre	$2(\text{near\_right\_border})$ if ((nearly_vertical_line > nearly_horizontal_line) and $(r \leq y/2)$) $2(1 - \text{near\_right\_border})$ if ((nearly_vertical_line > nearly_horizontal_line) and $(r > y/2)$) 0 otherwise

problems suitable fuzzy sets may be similarly synthesized from basic sets of fuzzy Hough transform. A non-null support of synthesized fuzzy set implies presence of corresponding feature in a pattern. The height of each synthesized fuzzy set is chosen to define feature element and set of n such feature elements constitute an n-dimensional feature vector for a character.

3.4 Genetic Algorithms for Feature Selection

In this section the feature selection problem in OCR systems is approached through GA. A number of neural network and fuzzy set theoretic approaches have been proposed for feature analysis in recent past [6]. A feature quality index (FQI) measure for ranking of features has been suggested by [11]. The feature ranking process is based on influence of feature on MLP output. It is related to the importance of feature in discriminating among classes. The impact of qth feature on MLP output out of a total of p features is measured by setting feature value to zero for each input pattern $x_i, i = 1, \ldots, n$. FQI is defined as the deviation of MLP output with qth feature value set to zero from output with all features present such that:

$$FQI_q = \frac{1}{n} \sum_{i=1}^{n} \left\| OV_i - OV_i^{(q)} \right\|^2 \tag{3.3}$$

In Eq. (3.3) OV_i and $OV_i^{(q)}$ are output vectors with all p features present and with qth feature set to zero. The features are ranked according to their importance as $q_1, \ldots, q_p$ if $FQI_{q_1} > \ldots > FQI_{q_p}$. In order to select best $p\prime$ features from the set of p features, $\binom{p}{p\prime}$ possible subsets are tested one at a time. The quality index $FQI_k^{(p\prime)}$ of kth subset S_k is measured as:

$$FQI_k^{(p\prime)} = \frac{1}{n} \sum_{i=1}^{n} \left\| OV_i - OV_i^k \right\|^2 \tag{3.4}$$

In Eq. (3.4) OV_i^k is MLP output vector with x_i^k as input where x_i^k is derived from x_i as:

$$x_{ij}^k = \begin{cases} 0 & \text{if } j \in S_k \\ x_{ij} & \text{ow} \end{cases} \tag{3.5}$$

A subset S_j is selected as optimal set of features if $FQI_j^{(p\prime)} \geq FQI_k^{(p\prime)} \forall k, k \neq j$. An important observation here is that value of $p\prime$ should be predetermined and $\binom{p}{p\prime}$ number of possible choices are verified to arrive at the best feature set. It is evident that no a priori knowledge is usually available to select the value $p\prime$ and an exhaustive search is to be made for all values $p\prime$ with $p\prime = 1, \ldots, p$. The number of possible trials is $(2^p - 1)$ which is prohibitively large for high values of p. To overcome drawbacks of above method, best feature set is selected by using of genetic algorithms [12]. Let us consider mask vector M where $M_i \in \{0, 1\}; i = 1, \ldots, p$ and each feature element $q_i, i = 1, \ldots, n$ is multiplied by corresponding mask vector element before reaching MLP input such that $I_i = q_i M_i$. MLP inputs are then written as:

$$I_i = \begin{cases} 0 & \text{if } M_i = 0 \\ q_i & \text{ow} \end{cases} \tag{3.6}$$

Thus, a particular feature q_i reaches MLP if corresponding mask element is one. To find sensitivity of a particular feature q_j, mask bit M_j is set to zero. With respect to the above discussions when kth subset of feature set $\{q_1, \ldots, q_p\}$ is selected, all corresponding mask bits are set to zero and rest are set to one. When feature set multiplied by these mask bits reaches MLP, the effect of setting features of subset S_k to zero is obtained. Then value of FQI_k is calculated. It is to be noted that kth subset thus chosen may contain any number of feature elements and not pre-specified $p\prime$ number of elements. Starting with an initial population of strings representing mask vectors, genetic algorithm is used with reproduction, crossover and mutation operators to determine best value of objective function. The objective function is FQI value of feature set S_k selected with mask bits set to zero for specific features and is given by:

$$FQI_k = \frac{1}{n} \sum_{i=1}^{n} \left\| OV_i - OV_i^k \right\|^2 \tag{3.7}$$

In this process, both the problems of predetermining value of $p\prime$ and searching through $\binom{p}{p\prime}$ possible combinations for each value of $p\prime$. In genetic algorithm implementation, the process is started with 20 features generated from fuzzy Hough transform such that the number of elements in mask vector is also 20. After running genetic algorithm for sufficiently large number of generations, mask string with best objective function value is determined. The feature elements corresponding to mask bits zero are chosen as selected set of features. The parameters for genetic algorithms are determined in terms of chromosome length, population size, mutation probability and crossover probability. MLP is next trained with only selected feature set for classification. The number of features varied when required. The genetic algorithm based feature selection method is shown in Fig. 3.6.

Fig. 3.6 The feature selection process using genetic algorithm

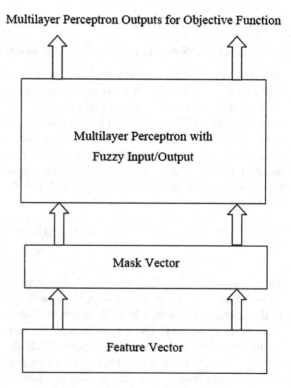

3.5 Rough Fuzzy Multilayer Perceptron

In this section RFMLP [7, 8, 17] for OCR systems is discussed. RFMLP is evolved from fuzzy MLP and rough sets. The methodology presented evolves RFMLP through modular concept using genetic algorithms to obtain a structured network for classification and rule extraction. The modular concept is based on divide and conquer strategy. It provides accelerated training and is a compact network which generates minimum number of rules with high certainty values. The variable mutation operator preserves the localized structure of knowledge based sub-networks as they are integrated and evolved. The dependency rules are generated from real valued attributes containing fuzzy membership values. For interested readers further details on rough sets are available in [19].

The MLP network consists of group of nodes arranged in layers. Each node in a layer is connected to all nodes in the next layer by links which have weight associated with them. The input layer contains nodes that represent input features of the classification problem. A real valued feature is represented by single node whereas discrete feature with n distinct values is represented by n input nodes. The classification strength of MLP is enhanced by incorporating rough and fuzzy sets. It has been used as an important soft computing tool for pattern classification [17]. MLP acts as efficient connectionist between the two. In this hybridization, fuzzy sets help in handling linguistic input information and ambiguity in output decision and rough sets extract domain knowledge for determining network parameters.

The fuzzy MLP [7, 8, 17] incorporates fuzziness at input and output levels of MLP. It is capable of handling both numerical as well as linguistic forms of input data. Any input feature is described in terms of some combination of membership values in linguistic property sets low (L), medium (M) and high (H). The class membership value μ of patterns is represented at output layer of fuzzy MLP. During training the weights are updated by backpropagation errors with respect to these membership values such that contribution of uncertain vectors is automatically reduced. A four layered feed forward MLP is used where output of a neuron in any layer h other than input layer ($h = 0$) is:

$$y_j^{(h)} = \frac{1}{1 + \exp\left(-\sum_i y_i^{(h-1)} w_{ji}^{(h-1)}\right)} \tag{3.8}$$

In Eq. (3.8), $y_i^{(h-1)}$ is the state of the ith neuron in preceding $(h-1)$th layer and $w_{ji}^{(h-1)}$ is the weight of connection from ith neuron in layer $(h-1)$ to jth neuron in layer h. For nodes in input layer y_j^0 corresponds to ith component of input vector. It is to be noted that $x_j^{(h)} = \sum_i y_i^{(h-1)} w_{ji}^{(h-1)}$. An n-dimensional pattern $F_i = [F_{i1}, F_{i2}, \ldots, F_{in}]$ is represented as $3n$-dimensional vector:

$$F_i = [\mu_{low(F_{i1})}(F_i), \ldots, \mu_{high(F_{in})}(F_i)] = [y_1^{(0)}, y_2^{(0)}, \ldots, y_{3n}^{(0)}] \tag{3.9}$$

In Eq. (3.9), μ values indicate membership functions of corresponding linguistic π-sets low, medium, and high along each feature axis and $y_1^{(0)}, y_2^{(0)}, \ldots, y_{3n}^{(0)}$ refer activations of $3n$ neurons in input layer. When input feature is exact in nature π-Fuzzy sets in one dimensional form are used with range [0, 1] and are represented as:

$$\pi(F_j; c, \lambda) = \begin{cases} 2\left(1 - \frac{||F_j - c||}{\lambda}\right)^2, for \lambda/2 \leq ||F_j - c|| \leq \lambda \\ 1 - 2\left(\frac{||F_j - c||}{\lambda}\right)^2, for 0 \leq ||F_j - c|| \leq \lambda/2 \\ 0, otherwise \end{cases} \quad (3.10)$$

In Eq. (3.10), $\lambda > 0$ is radius of π-function with c as the central point. Let us consider an l-class problem domain such that there are l nodes in output layer. Also consider n-dimensional vectors $o_k = [o_{k1}, \ldots, o_{kl}]$ and $v_k = [v_{k1}, \ldots, v_{kl}]$ which denote mean and standard deviation respectively of exact training data for kth class c_k. The weighted distance of training pattern F_i from kth class c_k is defined as:

$$z_{ik} = \sqrt{\sum_{j=1}^{n} \left[\frac{F_{ij} - o_{kj}}{v_{kj}}\right]^2} \text{ for } k = 1, \ldots, l \quad (3.11)$$

In Eq. (3.11), F_{ij} is value of jth component of ith pattern point. The membership of ith pattern in class k lying in range [0, 1] is:

$$\mu_k(F_i) = \frac{1}{1 + \left(\frac{z_{ik}}{f_d}\right)^{f_e}} \quad (3.12)$$

In Eq. (3.12), positive constants f_d and f_e are denominational and exponential fuzzy generators controlling amount of fuzziness in class membership set. RFMLP [7, 8, 17] works through five phases viz pre-processing, analysis, rule generation, classification and rule extraction and prediction. For sake of convenience analysis and rule generating phases are clubbed together.

In pre-processing phase decision table is created for rough set analysis. In this process data preparation tasks such as conversion, cleansing, completion checks, conditional attribute creation, decision attribute generation and attribute discretization are performed. Data splitting is performed which creates two randomly generated subsets, one subset containing objects for analysis and remaining subset containing objects for validation. It must be emphasized that data conversion performed on initial data must generate a form in which rough set tools can be applied.

The real world data often contain missing values. Since rough set classification involves mining for rules from data, objects with missing values in dataset may have undesirable effects on rules that are constructed. The data completion procedure removes all objects that have one or more missing values. Incomplete data in information systems exist broadly in practical data analysis. Approaches to complete the incomplete information system through various completion methods in pre-processing stage are normal in knowledge discovery process. However, these methods may lead to distortion in original data and knowledge. It can even render the original data to be unexplored. To overcome these deficiencies inherent in traditional methods, decomposition approach for incomplete information system i.e. decision table is used here [19].

Attributes in classification and prediction may have varying importance in problem domain. Their importance can be pre-assumed using auxiliary knowledge about the problem and expressed through proper choice of weights. However, when using rough sets for classification it avoids any additional information aside from what is included in information table itself. Basically rough sets try to determine from data available data in decision table whether all attributes are of same strength and if not how they differ in respect of classifier power. Therefore, some strategies for discretization of real value attributes are used when learning strategies for data classification with real value attributes are applied. The learning algorithm quality dependents on this strategy [9]. The discretization uses data transformation procedure which involves finding cuts in datasets that divide data into intervals. The values lying within an interval are then mapped to same value. This process leads to reduction in the size of attributes value set and ensures that rules that are mined are not too specific. For the discretization of continuous valued attributes rough sets with boolean reasoning [7, 8, 17] are considered here. This technique combines discretization of real valued attributes and classification. The algorithm is available in [17].

In analysis and rule generating phase the principal task is to compute reducts and corresponding rules with respect to particular kind of information system and decision system. The knowledge encoding is also embedded in this phase. We use relativised versions of matrices and functions viz d-reducts and d-discernibility matrices as basic tools for computation. The methodology is discussed as [17]:

Let $S = \langle U, A \rangle$ be a decision table with C and $D = \{d_1, \ldots, d_l\}$ as sets of condition and decision attributes respectively. We divide decision table $S = \langle U, A \rangle$ into l tables $S_i = \langle U_i, A_i \rangle, i = 1, \ldots, l$, corresponding to l decision attributes $d_1, \ldots, d_l$ where $U = U_1 \cup \ldots \cup U_l$ and $A_i = C \cup \{d_i\}$. Let $\{x_{i_1}, \ldots, x_{i_p}\}$ be the set of those objects of U_i that occur in $S_i, i = 1, \ldots, l$. Now for each d_i-reduct $B = \{b_1, \ldots, b_k\}$ discernibility matrix denoted by $M_{d_i}(B)$ from d_i-discernibility matrix is defined as:

$$c_{ij} = \{a \in B : a(x_i) \neq a(x_j)\} \quad for\ i, j = 1, \ldots, n \qquad (3.13)$$

For each object $x_j \in x_{i_1}, \ldots, x_{i_p}$ discernibility function $f_{d_j}^{x_j}$ is defined as:

$$f_{d_j}^{x_j} = \wedge \{\vee(c_{ij}) : 1 \leq i, j \leq n, j < i, c_{ij} \neq \varnothing\} \qquad (3.14)$$

In Eq. (3.14), $\vee(c_{ij})$ is disjunction of all members of c_{ij}. Then $f_{d_j}^{x_j}$ is brought to its conjunctive normal form. Thus dependency rule r_i is obtained such that $P_i \leftarrow d_i$ where P_i is disjunctive normal form $f_{d_j}^{x_j}, j \in i_1, \ldots, i_p$. The dependency factor df_i for r_i is:

$$df_i = \frac{card(POS_i(d_i))}{card(U_i)} \qquad (3.15)$$

Here $POS_i(d_i) = \bigcup_{X \in I_{d_i}} l_i(X)$ and $l_i(X)$ is lower approximation of X with respect to I_i. In this case $df_i = 1$. In knowledge encoding consider feature F_j for class c_k in l-class problem domain. The inputs for ith representative sample F_i are mapped to corresponding 3-dimensional feature space of $\mu_{low(F_{ij})}(F_i)$, $\mu_{medium(F_{ij})}(F_i)$, $\mu_{high(F_{ij})}(F_i)$ which correspond to low, medium and high values. Let us represented these by L_j, M_j, H_j respectively. As the method considers multiple objects in class, a separate $n_k \times 3n$-dimensional attribute value decision table is generated for each class c_k where n_k indicates number of objects in c_k. The absolute distance between each pair of objects is computed along each attribute L_j, M_j, H_j for all j. Equation (3.13) is modified to directly handle real valued attribute table consisting of fuzzy membership values. We define:

$$c_{ij} = \{a \in B : |a(x_i) - a(x_j)| > Th\} \text{ for } i, j = 1, \ldots, n_k \tag{3.16}$$

In Eq. (3.16), Th is an adaptive threshold. The adaptivity of this threshold is built in depending on inherent shape of membership function. While designing initial structure of RFMLP, the union of rules of l classes is considered. The input layer consists of $3n$ attribute values while output layer is represented by l classes. The hidden layer nodes model innermost operator in antecedent part of rule which can be either a conjunct or disjunct. The output layer nodes model outer level operands which can again be either a conjunct or disjunct. For each inner level operator corresponding to one output class, one dependency rule and one hidden node is dedicated. Only those inputs attribute that appear in this conjunct or disjunct are connected to appropriate hidden node which in turn is connected to corresponding output node. Each outer level operator is modeled at output layer by joining corresponding hidden nodes. It is noted that a single attribute involving no inner level operators is directly connected to appropriate output node via hidden node to maintain uniformity in rule mapping.

The classification phase [7, 8, 17] classifies rules generated from previous phase. Here the problem is effectively decomposed into sub-problems. As a result of this the problem is solved with compact networks and efficient combination as well as trained networks such that there is gain in terms of training time, network size and accuracy. The task is carried in two stages. In the first stage, l class classification problem is split into l two-class problems. Let there be l sets of sub-networks with $3n$ inputs and one output node each. The rough set theoretic concepts encode domain knowledge into each of the sub-networks using Eqs. (3.13)–(3.16). The number of hidden nodes and connectivity of knowledge based sub-networks is automatically determined. Each two-class problem leads to the generation of one or more crude sub-networks, each encoding a particular decision rule. Let each of these constitute pool knowledge based modules. Thus, $m \geq l$ such pools are obtained. Each pool k is perturbed to generate n_k sub-networks such that $n_1 = \cdots = n_k \cdots = n_m$. These pools constitute initial population of sub-networks which are then evolved independently using genetic algorithms. At the end of first stage modules or sub-networks corresponding to each two-class problem are concatenated to form an initial network for second stage. The inter module links are initialized to small random values as depicted in Fig. 3.7. A set of such concatenated networks forms initial population of genetic algorithms. The mutation probability

Fig. 3.7 Intra and inter-module links

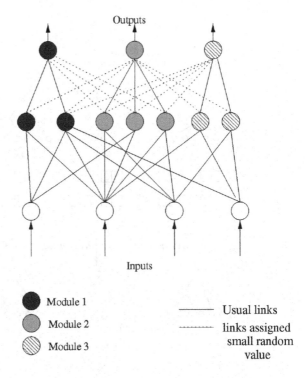

for inter-module links is now set to high value while that of intra-module links is set to relatively lower value. This restricted mutation helps preserve some of localized rule structures already extracted and are evolved as potential solutions. The initial population for genetic algorithms of entire network is formed from all possible combinations of these individual network modules and random perturbations about them. This ensures that for complex pattern distributions all different representative points remain in population. The algorithm then searches through reduced space of possible network topologies. The algorithm is available in [17].

Consider a problem of classifying a two dimensional data into two classes. The input fuzzy function maps features into 6-dimensional feature space. Let sample set of rules obtained from rough sets be:

$$c_1 \leftarrow (L_1 \wedge M_2) \vee (H_2 \wedge M_1); c_2 \leftarrow M_2 \vee H_1; c_3 \leftarrow L_3 \vee L_1 \qquad (3.17)$$

In Eq. (3.17), L_j, M_j, H_j correspond to $\mu_{low(F_j)}, \mu_{medium(F_j)}, \mu_{high(F_j)}$ respectively which denote low, medium and high values. For first phase of genetic algorithms three different pools are formed using one crude sub-network for class 1 and two crude sub-networks for class 2 respectively. Three partially trained sub-networks result from each of these pools. They are then concatenated to form $(1 \times 2 = 2)$ networks. The population for final phase of genetic algorithms is formed with these networks and perturbations about them. The steps followed in obtaining the final network structure is illustrated in Fig. 3.8.

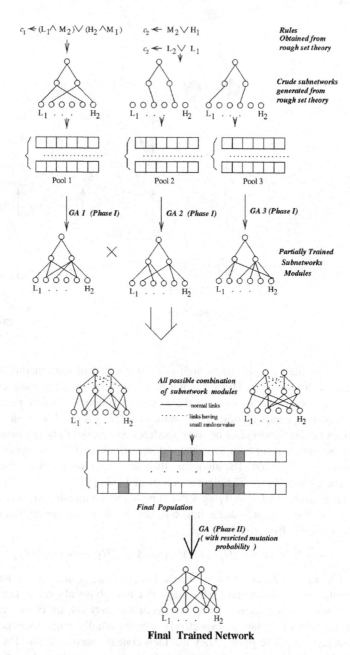

Fig. 3.8 Steps for designing a sample modular RFMLP

Now different features of genetic algorithm [12] relevant to this algorithm are highlighted. The chromosomal representation shown in Fig. 3.9. The problem variables consist of weight values and input/output fuzzy parameters. Each of the weights is encoded into a binary word of 16 bit length, where [000 . . . 0] decodes to -128 and [111 . . . 1] decodes to 128. An additional bit is assigned to each weight to indicate the presence or absence of link. The fuzzy parameters tuned are centers c and radius λ for each of the linguistic attributes low, medium and high of each feature and output fuzzy parameters f_d and f_e [2]. These are also coded as 16 bit strings in the range [0, 2]. For the input parameters, [000 . . . 0] decodes to 0 and [111 . . . 1] decodes to 1.2 times the maximum value attained by corresponding feature in the training set. The chromosome is obtained by concatenating all the above strings. Sample values of string length are around 2000 bits for reasonably sized networks.

The initial population is generated by coding networks obtained by rough set based knowledge encoding and by random perturbations about them. A population size of 64 is generally considered. It is obvious that due to large string length, single point crossover would have little effectiveness. Multiple point crossovers are adopted with distance between two crossover points being a random variable between 8 and 24 bits. This is done to ensure high probability for only one crossover point occurring within a word encoding a single weight. The search string being very large, the influence of mutation is more on search compared to crossover. The mutation probability has spatio-temporal variation. The maximum value of *pmut* is chosen to be 0.4 and minimum value as 0.01. The mutation probabilities also vary along encoded string. The bits corresponding to inter-module links being assigned probability *pmut* i.e. the value of *pmut* at that iteration and intra-module links assigned a probability *pmut*/10. This is done to ensure least alterations in the structure of individual modules already evolved. Hence, mutation operator indirectly incorporates domain knowledge extracted through rough sets. The objective function considered is of the form $F = \alpha_1 f_1 + \alpha_2 f_2$ where $f_1 =$ (No. of correctly classified sample in training set)/(Total no. of samples in training set); $f_2 = 1 - $ (No. of links present)/(Total no of links possible). Here α_1 and α_2 determine relative weight of each of the factors. The α_1 is taken as 0.9 and α_2 is taken as 0.1 to give more importance to classification score compared to network size in terms of number of links. The network connectivity, weights and input or

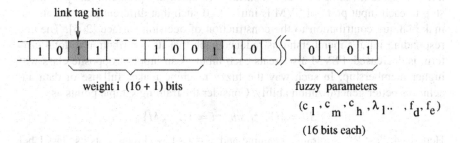

link tag bit

weight i (16 + 1) bits   fuzzy parameters

$(c_l, c_m, c_h, \lambda_l .. , f_d, f_e)$

(16 bits each)

Fig. 3.9 Chromosomal representation

output fuzzy parameters are optimized simultaneously. The selection is done by roulette wheel method. The probabilities are calculated on the basis of ranking of individuals in terms of objective function. Elitism is incorporated in selection process to prevent oscillation of fitness function with generation. The fitness of best individual of new generation is compared with current generation. If latter has higher value, the corresponding individual replaces randomly selected individual in new population.

In rule extraction and prediction phase rules from previous phases are extracted and utilizes them for recognition task. The algorithm for rule extraction considered here is decomposition based and is adopted from [17]. Since training algorithm imposes a structure on network which results in a sparse network having few strong links such that threshold values are well separated. Hence, rule extraction algorithm generates most of the embedded rules over small number of computational steps. An important consideration is the order of application of rules in rule base. Since most of the real life patterns are noisy and overlapping, rule bases obtained are often not totally consistent. Hence, multiple rules may fire for a single example. Several existing approaches apply the rules sequentially [6] often leading to degraded performance. The rules extracted have confidence factors associated with them. Thus if multiple rules are fired, the strongest rule having the highest confidence is chosen.

3.6 Fuzzy and Fuzzy Rough Support Vector Machines

In this section two important versions of SVM viz FSVM [3, 4] and FRSVM [2, 5] for OCR systems are presented. FSVM is developed by introducing the concept of fuzzy membership function [27] to SVM. Several versions of SVM are available in the literature [6]. In classical SVM each sample point is treated equally i.e. each input is fully assigned to one of the two classes as shown in Fig. 3.10. The separating hyperplane between classes leads to different support vectors [6]. For interested readers further details are available in [2–5].

However, in many applications some input points are detected as outliers and may not be exactly assigned to one of the two classes. Here each point does not have the same meaning to decision surface. To solve this problem, fuzzy membership to each input point of SVM is introduced such that different input points can make unique contribution to the construction of decision surface [2–5]. The corresponding input's membership is reduced such that its contribution to total error term is decreased. FSVM also treats each input as an input of opposite class with higher membership. In such way the fuzzy machine makes full use of data and achieves better generalization ability. Consider the training sample points as:

$$SP = \{(Y_i, z_i, sm_i); i = 1, \dots, M\} \tag{3.18}$$

Here, each $Y_i \in R^N$ is a training sample and $z_i \in \{-1, +1\}$ represents its class label; $sm_i; i = 1, \dots, M$ is fuzzy membership which satisfies $s_j \le sm_i \le s_i; i, j = 1, \dots, M$

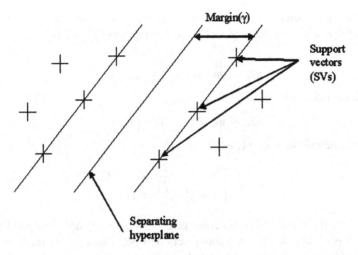

Fig. 3.10 The separating hyperplane between classes leading to different support vectors

with sufficiently small constant $s_j > 0$ and $s_i \leq 1$. The set $P = \{Y_i | (Y_i, z_i, sm_i) \in SP\}$ thus contains two classes. One class contains sample point Y_i with $z_i = 1$ denoted by C^+ such that:

$$C^+ = \{Y_i | Y_i \in SP \wedge z_i = 1\} \qquad (3.19)$$

Other class contains sample point Y_i with $z_i = -1$ denoted by C^- such that:

$$C^- = \{Y_i | Y_i \in SP \wedge z_i = -1\} \qquad (3.20)$$

It is obvious that $P = C^+ \cup C^-$. The quadratic classification problem can be represented as:

$$\min \frac{1}{2} \|w\|^2 + C \sum_{i=1}^{M} sm_i \xi_i \qquad (3.21)$$

$$\text{subject to}: \begin{cases} z_i(w^T \Phi(Y_i) + b) \geq 1 - \xi_i, i = 1, \dots, M \\ \xi_i \geq 0 \end{cases}$$

In Eq. (3.21), C is constant. Since the fuzzy membership sm_i governs behavior of corresponding sample point Y_i towards one class and parameter ξ_i is the error measure in SVM. The term $sm_i \xi_i$ can be considered as error measure with different weights. A smaller sm_i reduces the effect of parameter ξ_i in Eq. (3.21) such that corresponding point Y_i can be treated as less significant. The above quadratic problem can also be solved by their corresponding dual problems [1]. The choice of appropriate fuzzy memberships is vital towards the success of FSVM. The fuzzy membership function for reducing the effect of outliers is function of distance between each data point and its corresponding class center is represented through input space parameters. Given sequence of training points in Eq. (3.21) the mean of class C^+ and C^- are denoted as Y_+ and Y_- respectively. The radius of class C^+ is:

$$rd_+ = \max\|Y_+ - Y\|_i; \, Y_i \in C^+ \tag{3.22}$$

The radius of class C^- is:

$$rd_- = \max\|Y_- - Y_i\|; \, Y_i \in C^- \tag{3.23}$$

The fuzzy membership fm_i [1] is:

$$fm_i = \begin{cases} 1 - \frac{\|Y_+ - Y_i\|}{(rd_+ + \varepsilon)} & if \, Y_i \in C^+ \\ 1 - \frac{\|Y_- - Y_i\|}{(rd_- + \varepsilon)} & if \, Y_i \in C^- \end{cases} \tag{3.24}$$

FSVM with the membership function given in Eq. (3.24) achieves good performance. A particular sample in training set contributes little to the final result and outliers' effect of outliers are eliminated by taking average on samples.

An enhanced version of FSVM viz MFSVM [3] is also discussed here. Consider a training sample $Y_i \in S$. Let $\Phi(Y_i)$ be a mapping function from input into feature space. The kernel function used is hyperbolic tangent kernel given by $K(Y_i, Y_j) = \tanh\left[\Phi(Y_i) \cdot \Phi(Y_j)\right]$ and shown in Fig. 3.11 [3]. It is known as sigmoid kernel and has its roots in artificial neural networks (ANN) where bipolar sigmoid function is used as activation function for artificial neurons. MFSVM using hyperbolic tangent kernel function is equivalent to two layer perceptron ANN. It is conditionally positive definite and performs well in practice. This kernel allows lower computational cost and higher rate of positive eigenvalues of kernel matrix which alleviates limitations of other kernels.

It was pointed out that kernel matrix of hyperbolic tangent function is not positive semi definite for certain parameters. More discussions are available in [3]. Generally when kernel is not positive semi definite, the function cannot be satisfied and primal dual relationship does not exist. Thus, it becomes quite unclear

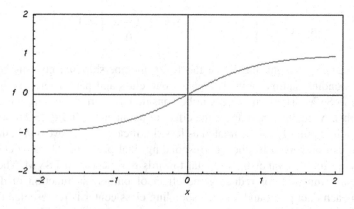

Fig. 3.11 The graphic representation of hyperbolic tangent kernel for real values

what type of classification problems is to be solved. The sigmoid kernel has been used in several applications with appreciable success [3]. This fact motivates hyperbolic tangent kernel usage for defining fuzzy membership function in MFSVM. Define Φ_+ as the class center of C^+ in feature space as:

$$\Phi_+ = \frac{1}{m_+} \sum_{Y_i \in C^+} \Phi(Y_i) f \qquad (3.25)$$

In Eq. (3.25), m_+ is number of samples of class C^+ with f_i the frequency of ith sample in feature space $\Phi(Y_i)$. Again Φ_- is class center of C^- in feature space is defined as:

$$\Phi_- = \frac{1}{m_-} \sum_{Y_i \in C^-} \Phi(Y_i) f_i \qquad (3.26)$$

In Eq. (3.26), m_- is number of samples of class C^- with f_i the frequency of ith sample in feature space $\Phi(Y_i)$. The radius of C^+ is:

$$rd_+ = \frac{1}{n} \max \| \Phi_+ - \Phi(Y_i) \| \qquad (3.27)$$

Here $Y_i \in C^+$ and radius of C^- is:

$$rd_- = \frac{1}{n} \max \| \Phi_- - \Phi(Y_i) \| \qquad (3.28)$$

Here $Y_i \in C^-$ and $n = \sum_i f_i$.
Then

$$
\begin{aligned}
rd_+^2 &= \frac{1}{n} \max \left\| \Phi(Y') - \Phi_+ \right\|^2 = \frac{1}{n} \max \left[\Phi^2(Y') - 2\Phi(Y') \cdot \Phi_+ + \Phi_+^2 \right] \\
&= \frac{1}{n} \max \left[\Phi^2(Y') - \frac{2}{m_+} \sum_{Y_i \in C^+} tanh \left[\Phi(Y_i) \cdot \Phi(Y') \right] + \frac{1}{m_+^2} \sum_{Y_i \in C^+} \sum_{Y_j \in C^+} tanh \left[\Phi(Y_i) \cdot \Phi(Y_j) \right] \right] \\
&= \frac{1}{n} \max \left[K(Y', Y') - \frac{2}{m_+} \sum_{Y_i \in C^+} K(Y_i, Y') + \frac{1}{m_+^2} \sum_{Y_i \in C^+} \sum_{Y_j \in C^+} K(Y_i, Y_j) \right]
\end{aligned}
$$

In Eq. (3.27) $Y' \in C^+$ and m_+ is number of training samples in C_+.

$$(3.29)$$

Similarly,

$$rd_-^2 = \frac{1}{n} \max \left[K(Y', Y') - \frac{2}{m_-} \sum_{Y_i \in C^-} K(Y_i, Y') + \frac{1}{m_-^2} \sum_{Y_i \in C^-} \sum_{Y_j \in C^-} K(Y_i, Y_j) \right]$$

$$(3.30)$$

In Eq. (3.30), $Y' \in C^-$ and m_- is number of training samples in C_-. The square of distance between sample $Y_i \in C^+$ and its class center in feature space is:

$$dist_{i+}^2 = \| \Phi(Y_i) - \Phi_+ \|^2 = \Phi^2(Y_i) - 2\tanh \left[\Phi(Y_i) \cdot \Phi_+ \right] + \Phi_+^2$$

$$dist_{i+}^2 = K(Y_i, Y_j) - \frac{2}{m_+} \sum_{Y_j \in C^+} K(Y_i, Y_j) + \frac{1}{m_+^2} \sum_{Y_j \in C^+} \sum_{Y_k \in C^+} K(Y_j, Y_k)$$

(3.31)

Similarly the square of distance between sample $Y_i \in C^-$ and its class center in feature space is:

$$dist_-^2 = K(Y_i, Y_j) - \frac{2}{m_-} \sum_{Y_j \in C^-} K(Y_i, Y_j) + \frac{1}{m_-^2} \sum_{Y_j \in C^-} \sum_{Y_k \in C^-} K(Y_j, Y_k)$$

(3.32)

Now $\forall i; i = 1, \ldots \ldots, M$ fuzzy membership function fm_i [3] is:

$$fm_i = \begin{cases} 1 - \sqrt{\dfrac{\left\| dist_{i+}^2 \right\| - \left\| dist_{i+}^2 \right\| \cdot rd_+^2 + rd_+^2}{\left(\left\| dist_{i+}^2 \right\| + \left\| dist_{i+}^2 \right\| \cdot rd_+^2 + rd_+^2 \right) + \varepsilon}} & if \ z_i = 1 \\[3ex] 1 - \sqrt{\dfrac{\left\| dist_{i-}^2 \right\| - \left\| dist_{i-}^2 \right\| \cdot rd_-^2 + rd_-^2}{\left(\left\| dist_{i-}^2 \right\| + \left\| dist_{i-}^2 \right\| \cdot rd_-^2 + rd_-^2 \right) + \varepsilon}} & if \ z_i = -1 \end{cases}$$

(3.33)

In Eq. (3.33), $\varepsilon > 0$ such that sm_i is never zero. The membership function is function of center and radius of each class in feature space and is represented with kernel. The training samples used here can be either linear or nonlinear separable. In former separating surface can be computed in input space. In later input space should be mapped into high dimensional feature space to compute separating surface using linear separating method.

In formulation of FSVM fuzzy membership reduces outliers' effects. It is calculated in input space whenever training samples are linear or nonlinear separable. When samples are nonlinear separable fuzzy memberships are calculated in input space but not in feature space. The contribution of each point in constructing hyperplane in feature space cannot be represented properly. The fuzzy membership function illustrated in Eq. (3.33) efficiently solves this problem. By representing fuzzy membership with mapping function the input space is mapped into feature space. The fuzzy memberships can be calculated in feature space. Further using kernel function it is not required to know the shape of mapping function. This method more accurately represents contribution of each sample point towards constructing separating hyperplane in feature space [6]. Thus MFSVM reduces effect of outliers more efficiently and has better generalization ability. The performance of fuzzy membership given in Eq. (3.33) can be improved through rough fuzzy membership function. For $\forall i; i = 1, \ldots, M$ fuzzy rough membership function $frm_i(p)$ is [2]:

$$frm_i(p) = \begin{cases} 1 - \left(\dfrac{\sum_{i=1}^{H} \mu_{FC_i}(p)\tau_{C_c}^i}{\sum_i \mu_{FC_i}(p)} \right) \sqrt{\dfrac{\left| \|dist_{i+}^2\| - \|dist_{i+}^2\| \right| \cdot rd_+^2 + rd_+^2}{\left(\|dist_{i+}^2\| + \|dist_{i+}^2\| \right) \cdot rd_+^2 + rd_+^2) + \varepsilon}} & if \ z_i = 1 \\[3ex] 1 - \left(\dfrac{\sum_{i=1}^{H} \mu_{FC_i}(p)\tau_{C_c}^i}{\sum_i \mu_{FC_i}(p)} \right) \sqrt{\dfrac{\left| \|dist_{i-}^2\| - \|dist_{i-}^2\| \right| \cdot rd_-^2 + rd_-^2}{\left(\|dist_{i-}^2\| + \|dist_{i-}^2\| \right) \cdot rd_-^2 + rd_-^2) + \varepsilon}} & if \ z_i = -1 \end{cases}$$

$$(3.34)$$

The term $(\cdot)$ in Eq. (3.34) holds when $(\exists i)\mu_{FC_i}(p) > 0$ and $\varepsilon > 0$ so that $frm_i(p) \neq 0$. Here $\tau_{C_c}^i = \dfrac{\|FC_i \cap C_c\|}{\|FC_i\|}$ and $\dfrac{1}{\sum_i \mu_{FC_i}(p)}$ normalizes fuzzy rough membership function $\mu_{FC_i}(p)$. *The function is constrained fuzzy rough membership function* [2]. The above definition can be modified as [2]:

$$frm_i^c(p) = \begin{cases} 1 - \left(\dfrac{\sum_{i=1}^{H} \mu_{FC_i}(p)\tau_{C_c}^i}{\widehat{H}} \right) \sqrt{\dfrac{\left| \|dist_{i+}^2\| - \|dist_{i+}^2\| \right| \cdot rd_+^2 + rd_+^2}{\left(\|dist_{i+}^2\| + \|dist_{i+}^2\| \right) \cdot rd_+^2 + rd_+^2) + \varepsilon}} & if \ z_i = 1 \\[3ex] 0 & if \ z_i = -1 \end{cases}$$

$$(3.35)$$

In Eq. (3.35) $\widehat{H}$ is number of classes such that p has non-zero membership. When p does not belong to any class then $\widehat{H} = 0$ as a result of which $\dfrac{\sum_{i=1}^{H} \mu_{FC_i}(p)\tau_{C_c}^i}{\widehat{H}}$ becomes undefined. This issue is resolved by taking $frm_i^c(p) = 0$ when p does not belong to any class. This definition does not normalize fuzzy rough membership values and so the function is possibilistic fuzzy rough membership function. Equations (3.34) and (3.35) express the fact that if an input pattern belongs to class i.e. all belonging to only one class with non-zero memberships then no fuzzy roughness are involved. However in above equation it matters to what extent the pattern belongs to classes.

The fuzzy rough membership values depend on fuzzy classification of input dataset. The fuzziness in classes represents fuzzy linguistic uncertainty present in dataset. The classification can be performed through either (a) unsupervised classification which involves collecting data from all classes and classify them subsequently without considering associated class labels with data or (b) supervised classification where separate datasets are formed for each class and classification is performed on each such dataset to find subgroups present in data from same class. Both classification tasks can be performed by some trivial classification algorithms [2]. However, there are certain problems which are to be taken care of such as: (a) number of classes which have to be fixed a priori or which may not be known (b) it will not work in case number of class is one and (c) generated fuzzy memberships are not possibilistic.

To overcome the first problem evolutionary programming based method may be used [2]. For various classification problems evolutionary methods can automatically determine number of classes. It is worth mentioning that number of classes should be determined as best as possible. Otherwise calculation of fuzzy linguistic variables will be different and as a result fuzzy rough membership values may also vary. For second problem if it is known a priori that only one class is present then mean and standard deviation are calculated from input dataset and π fuzzy membership curve is fitted. But while doing so care must be taken to detect possible presence of the outliers in input dataset. To overcome third problem possibilistic fuzzy classification algorithm or any mixed classification algorithm can be used. As of now there is no single classification algorithm which can solve all the problems. If output class is fuzzy then it may be possible to assign fuzzy memberships for output class subjectively. However, if domain specific knowledge is absent then we have to be satisfied with given crisp membership values.

The fuzzy rough ambiguity plays critical role in many classification problems because of its capability towards modeling non statistical uncertainty. The characterization and quantification of fuzzy roughness are important aspects affecting management of uncertainty in classifier design. Hence measures of fuzzy roughness are essential to estimate average ambiguity in output class. A measure of fuzzy roughness for discrete output class $C_c \subseteq X$ is a mapping $S(X) \rightarrow \Re^+$ that quantifies degree of fuzzy roughness present in C_c. Here $S(X)$ is set of all fuzzy rough power sets defined within universal set X. The fuzzy rough ambiguity must be zero when there is no ambiguity in deciding whether an input pattern belongs to it or not. The equivalent classes form fuzzy classes so that each class is fuzzy linguistic variable. The membership is function of center and radius of each class in feature space and is represented with kernel.

Here fuzzy membership reduces outliers' effects [2]. When samples are nonlinear separable fuzzy memberships are calculated in input space but not in feature space. The contribution of each point in hyperplane in feature space cannot be represented properly and fuzzy rough membership function efficiently solves this. Through fuzzy rough membership function the input is mapped into feature space. The fuzzy rough memberships are calculated in feature space. Further using kernel function it is not required to know shape of mapping function. This method represents contribution of each sample point towards separating hyperplane in feature space. The proposed machine reduces outlier' effects efficiently and has better generalization ability.

The higher value of fuzzy rough membership function implies importance of data point to discriminate between classes [2]. It implies highest value is given by support vectors. These vectors are training points which are not classified with confidence. These are examples whose corresponding α_i values are non-zero. From representer theorem [27] optimal weight vector w^* is linear combination of support vectors which are essential training points. The number n_{SV} of support vectors also characterizes complexity of learning task. If n_{SV} is small then only a few examples are important and rest can be disregarded. If n_{SV} is large then nearly every example is important for accuracy. It has been shown that under general assumptions about loss function and underlying distribution training data $n_{SV} = \Omega(n)$. This suggests that asymptotically all points are critical for training. While this gives $\Omega(n)$ bound on training time this solves classification problem exactly. Further datasets need not necessarily have $\Theta(n)$ support vectors.

3.7 Hierarchical Fuzzy Bidirectional Recurrent Neural Networks

In this section hierarchical fuzzy version of bidirectional recurrent neural networks viz HFBRNN for OCR systems is presented. HFBRNN [1] takes full advantage of deep recurrent neural network (RNN) towards modeling long-term information of data sequences. The recognition of characters done by HFBRNN at review level. The performance of HFBRNN is improved by fine tuning parameters of the network in a hierarchical fashion. The motivation is obtained from long short term memory (LSTM) [1] and bidirectional LSTM (BLSTM) [1]. The evaluation is done on different types of highly biased character data. For interested readers further details are available in [1].

RNNs are deep learning ANNs [1] where connections between different computational units form directed cycle. This creates an internal network state that exhibits its dynamic temporal behavior. RNNs use internal memory to process arbitrary input sequences. This makes them suitable for non-segmented OCR tasks. RNNs are more efficient than traditional ANNs and SVMs [6] because they can be trained in either supervised or unsupervised manner. The network learns something intrinsic about data without help of target vector and is stored as network weights. The unsupervised training in network has identical input as target units. In deep learning optimization routine applied to network architecture itself. The network is directed graph where each hidden unit is connected to other hidden units. Each hidden layer going further into network is non-linear combination of layers because of combination of outputs from all previous units' with their activation functions. When optimization routine is applied to network, each hidden layer becomes optimally weighted and non-linear layer. When each sequential hidden layer has fewer units than one below it then each hidden layer becomes low dimensional projection of layer below it. With recurrent structure, RNN models contextual information of temporal sequence. Generally it is very difficult to train RNNs with commonly used activation functions due to vanishing gradient and error blowing up problems [1]. To solve this LSTM architecture is used [1] which replaces nonlinear units in traditional RNNs. The Fig. 3.12 illustrates LSTM memory block with single cell. It contains one self-connected memory cell and three multiplicative units viz input gate, forget gate and output gate which can store and access long range contextual information of temporal sequence. The activations of memory cell and three gates are available in [1]. In order to utilize past and future context, BRNN is used through forward and backward sequence [1] to two separate recurrent hidden layers. These two recurrent hidden layers share same output layer.

The underlying fuzzy membership function which takes care of the inherent vagueness and impreciseness is represented through trapezoidal fuzzy numbers [27]. The trapezoidal membership function is represented through Eq. (3.36) and

Fig. 3.12 Long short term
memory block with one cell

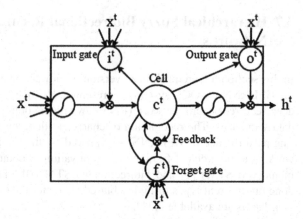

Fig. 3.13 The trapezoidal
membership function defined
by trapezoid $(x; a, b, c, d)$

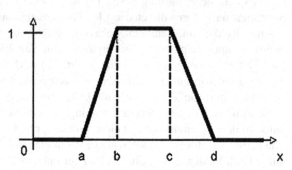

Fig. 3.13 is a piecewise linear, continuous function defined within interval $[0,1]$
and controlled by four parameters a, b, c, d and $x \in \mathbb{R}$ [27]:

$$\mu_{trapezoid}(x; a, b, c, d) = \begin{cases} 0, & x \leq a \\ \frac{x-a}{b-a}, & a \leq x \leq b \\ 1, & b \leq x \leq c \\ \frac{d-x}{d-c}, & c \leq x \leq d \\ 0, & d \leq x \end{cases} \qquad (3.36)$$

Now BRNN [1] is evaluated in terms of RNN and trapezoidal fuzzy membership
function [27] which leads to the development of HFBRNN. Instead of providing
output for each type of character, the model gives only outputs as final prediction.
To capture the entire context, backpropagation-through-time parameter is selected
so that it exceeds character length. The different character sets are expressed
through various labels with several recognition aspects [1]. For each of these recognition aspects there is—1, 0, 1 such that there is one-hot vector of 3 elements
for each;—1 (most negative), 1 (most positive) and 0 (neutral). If the character does not mention any recognition aspect it is assumed neutral. For different

recognition aspects prediction becomes $\hat{z} \in R^{21}$. Considering $y^{(1)}, y^{(2)}, \ldots \ldots$ the forward propagation is:

$$p^{(t)} = \sigma\left(W_p p^{(t-1)} + W_y y^{(t)}\right) \tag{3.37}$$

The final output for each recognition aspect results in:

$$\hat{z} = softmax\left(W_s h^{t=T}\right) \tag{3.38}$$

Here is concatenation of single predictions for each recognition aspect of the product:

$$\hat{z} = \left(\hat{z}_1 \; \hat{z}_2 \; \hat{z}_3 \; \hat{z}_4 \; \hat{z}_5 \; \hat{z}_6 \; \hat{z}_7\right)^T \tag{3.39}$$

This leads to the correct recognition character at the end. The matrices W_y, W_p, W_s and M are character vectors which are required to be learned. The idea behind this structure is that RNNs accumulate characters over the whole context. The post character context is not considered as the character is observed only in one direction. In order to determine character recognition aspect FBRNN is used. In FBRNN accumulation task is performed in two directions which allows more flexibility. The model runs through the sequence of characters in reverse order with different set of parameters that are updated. In order to specify backward channel sequence of characters are inverted and the same RNN is performed as done before on other direction. The final output is calculated concatenating p_g and p_h from both the directions:

$$p_g^{(t)} = \sigma\left(W_{p_g} p_g^{(t-1)} + W_y y^{(t)}\right) \tag{3.40}$$

$$p_h^{(t)} = \sigma\left(W_{p_h} p_h^{(t-1)} + W_y y_{inverted}^{(t)}\right) \tag{3.41}$$

$$\hat{z} = softmax\left(W_{s,brnn}\begin{pmatrix} p_g \\ p_h \end{pmatrix} + b_s\right) \tag{3.42}$$

In order to capture recognition aspects context in more granular way LSTM version of RNN is deployed here. Instead of just scanning character sequence in order the model stores information in gated units in an input gate $i^{(t)}$ with weight on current cell, a forget gate $f^{(t)}$, an output gate $o^{(t)}$ to specify relevance of current cell content and new memory cell $\tilde{cc}^{(t)}$. For time series tasks of unknown length LSTM are capable of storing and forgetting information better than their counterparts [1].

$$i^{(t)} = \sigma\left(W_i y^t + V_p p^{(t-1)}\right) \tag{3.43}$$

$$f^{(t)} = \sigma\left(W_f y^t + V_f p^{(t-1)}\right) \tag{3.44}$$

$$o^{(t)} = \sigma\left(W_o y^t + V_o p^{(t-1)}\right) \tag{3.45}$$

$$\widetilde{cc}^{(t)} = tanh\left(W_{cc} y^t + V_{cc} p^{(t-1)}\right) \tag{3.46}$$

$$cc^{(t)} = f^{(t)} cc^{(t-1)} + i^{(t)} \widetilde{cc}^{(t)} \tag{3.47}$$

$$p^{(t)} = ot^{(t)} tanh\left(cc^{(t)}\right) \tag{3.48}$$

The prediction now becomes:

$$\hat{z} = softmax(W_z p + b_z) \tag{3.49}$$

The bidirectional LSTM version of RNN scans sequence of characters in reverse order using second set of parameters. The final output is concatenation of final hidden vectors from original and reversed sequence:

$$\hat{z} = softmax\left(W_z\begin{pmatrix}p_g^T \\ p_h^T\end{pmatrix} + b_z\right) \tag{3.50}$$

The standard version of RNN performs below expectations as most reviews do not contain detectable recognition aspects with positive or negative values. The prior distribution of dataset is generally biased towards 0 class (neutral class). The model tends to always predict 0 and is not capable to predict −1 or 1.

Finally, the hierarchical version of BRNN viz HBRNN [1] is presented in terms of BRNN. The computational benefits received from BRNN serve the major motivation. HBRNN is different from BRNN in terms of efficient classification accuracy based on similarities and running time when volume of data grows [1]. The architecture of the proposed model is shown in Fig. 3.14 where temporal sequences in review are modeled by BRNNs which are combined together to form HBRNN. The model is composed of 6 layers viz $br_1 - br_2 - br_3 - br_4 - fc - sm$. Here, br_i; $i = 1, 2, 3, 4$ denote layers with BRNN nodes, fc denotes fully connected layer and sm denotes softmax layer. In HBRNN each layer takes care of classification tasks [1] and plays a vital role in success of whole network. Each layer constitutes hierarchy of classifier. To recover any single hierarchy split BRNN is run on small subset of character data to compute seed classification value. The subset of input dataset is produced randomly. This activity starts at layer br_1. Using the initial classification value, remaining character data is placed into seed class for which it is most similar on average. This results in classification of entire dataset using only similarities to characters in small subset. By recursively applying this procedure to each class HBRNN is obtained using small fraction of similarities. The classification task proceeds till br_4. In this recursive

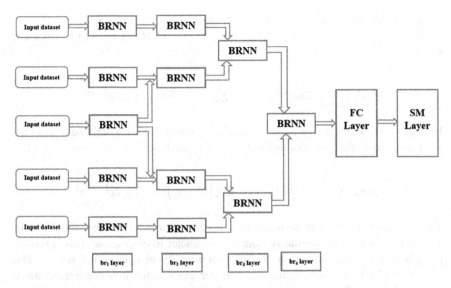

Fig. 3.14 The architecture of proposed HBRNN model

phase no measurements are observed between classes at previous split. This results in robust HBRNN that aligns its measurements mt to resolve higher resolution in the class structure. The pseudo code for HBRNN is shown in algorithm below.

Algorithm: HBRNN $\left(BRNN, mt, \{y_i\}_{i=1}^{Wr_j}, Cs_j \right)$

if $Wr_j < mt$ **then return** $\{y_i\}_{i=1}^{Wr_j}$

Select $V \subseteq \{y_i\}_{i=1}^{Wr_j}$ of size v uniformly at random

$C_1', \ldots \ldots, C_{Cs_j}' \leftarrow BRNN(V, Cs_j)$

Set $C_1 \leftarrow C_1', \ldots \ldots, C_{Wr_j} \leftarrow C_{Wr_j}'$

for $y_i \in \{y_i\}_{i=1}^{Wr_j} \setminus V$ **do**

$\forall k \in [Cs_j], \alpha_k \leftarrow \frac{1}{|C_j'|} \sum_{y_s \in C_j'} S(y_i, y_s)$

$C_{\text{argmax}_{k \in [Cs_j]} \alpha_k} \leftarrow C_{\text{argmax}_{k \in [Cs_j]} \alpha_i} \cup \{y_i\}$

end for

output $\{C_k, HBRNN \left(BRNN, mt, C_k, Cs_j \right) \}_{j=1}^{Cs_j}$

HBRNN is characterized in terms of probability of success in recovering true hierarchy Cs^*, measurement and runtime complexity. Some restrictions are placed on similarity function S such that similarities agree with hierarchy up to some random noise:

S1 For each $y_i \in Cs_j \in Cs^*$ and $j\prime \neq j$:

$$\min_{y_p \in Cs_j} \text{Exp}[S(y_i, y_p)] - \max_{y_p \in Cs_j} \text{Exp}[S(y_i, y_p)] \geq \gamma > 0$$

Here expectations are taken with respect to the possible noise on S.

S2 For each $y_i \in Ct_j$, a set of V_j words of size v_j drawn uniformly from Cs_j satisfies:

$$\mathbb{Prob}\left(\min_{y_p \in Cs_j} \mathbb{Exp}[S(y_i, y_p)] - \sum_{y_p \in V_j} \frac{S(y_i, y_p)}{v_j} > \epsilon \right) \leq 2e^{\left\{\frac{-2v_j\epsilon^2}{\sigma^2}\right\}}$$

Here $\sigma^2 \geq 0$ parameterizes noise on similarity function S. Similarly set $V_{j'}$ of size $v_{j'}$ drawn uniformly from cluster $Cs_{j'}$ with $j \neq j$ satisfies:

$$\mathbb{Prob}\left(\sum_{y_p \in V_{j'}} \frac{S(y_i, y_p)}{v_{j'}} - \max_{y_p \in C_{j'}} \mathbb{Exp}[S(y_i, y_p)] > \epsilon \right) \leq 2e^{\left\{\frac{-2v_{j'}\epsilon^2}{\sigma^2}\right\}}$$

The condition S1 states that similarity from character y_i to its class should be in expectation larger than similarity from that character to other class. This is related to tight classification condition [1] and is less stringent than earlier results. The condition S2 enforces that within-and-between-class similarities concentrate away from each other. This condition is satisfied if similarities are constant in expectation perturbed with any subgaussian noise. From the viewpoint of feature learning stacked BRNNs extracts temporal features of sentiment sequences in data. After obtaining features of character sequence, fully connected layer fc and softmax layer sm performs classification. The LSTM architecture effectively overcomes vanishing gradient problem [1]. The LSTM neurons are adopted in last recurrent layer $br4$. The first three BRNN layers use tanh activation function. This is trade-off between improving representation ability and avoiding over fitting. The number of weights in LSTM is more than that in tanh neuron. It is easy to overfit network with limited character training sequences. The algorithm has certain shortcomings for practical applications. Specifically if Cs is known and constant across splits in hierarchy, above assumptions are violated in practice. This is resolved by fine tuning the algorithm with heuristics. The eigengap is employed where Cs is chosen such that eigenvalues gap of laplacian is large. All subsampled characters are discarded with low degree when restricted to sample with removes underrepresented classes from sample. In averaging phase if character are not similar to any represented class, new class for the character is created.

3.8 Fuzzy Markov Random Fields

In this section fuzzy markov random fields is presented. The motivation of fuzzy markov random fields is adopted from [26] where type-2 (T2) fuzzy sets as shown in Fig. 3.15 are integrated with markov random fields (MRFs) resulting in T2 FMRFs. The T2 FMRFs handles both fuzziness and randomness in the structural pattern representation. The T2 membership function has a 3–D structure in which the primary membership function describes randomness and the secondary

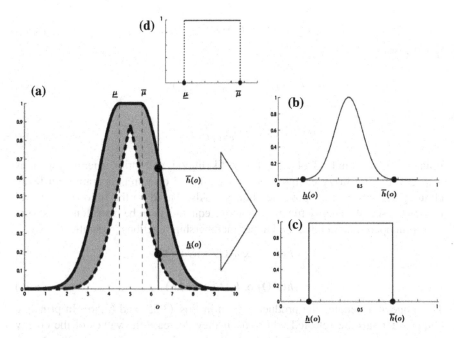

Fig. 3.15 Three-dimensional type-2 fuzzy membership function **a** primary membership with (*thick dashed line*) lower and (*thick solid line*) upper membership functions where $\underline{h}$(o) and $\bar{h}$(o) are lower and upper bounds of the primary membership of the observation o; the shaded region is the foot print of uncertainty **b** gaussian secondary membership function **c** interval secondary membership function **d** mean μ has a uniform membership function

membership function evaluates the fuzziness of the primary membership function. The MRFs represent statistical patterns structurally in terms of neighborhood system and clique potentials [26] and have been widely applied to image analysis and computer vision. The T2 FMRFs defines the same neighborhood system as the classical MRFs. In order to describe the uncertain structural information in patterns, the fuzzy likelihood clique potentials are derived from T2 fuzzy Gaussian mixture models. The fuzzy prior clique potentials are penalties for the mismatched structures based on prior knowledge. The T2 FMRFs models the hierarchical character structures present in the language characters. For interested readers further details are available in [26].

Basically the hidden fuzzy MRFs [26] characterize fuzzy image pixels that are difficult to segment. The T2 FMRF uses fuzzy clique potentials to evaluate the uncertain labeling configuration at all pixels. By integrating T2 fuzzy sets the estimation in terms of fuzzy posterior energy function is:

$$\mathcal{F}^* = \underbrace{\arg\min}_{\mathcal{F}} U_{\bar{\lambda}}(\mathcal{F}O) \tag{3.51}$$

$$U_{\bar{\lambda}}(\mathcal{F}|O) \propto U_{\bar{\lambda}}(O|\mathcal{F}) \sqcup U_{\bar{\lambda}}(\mathcal{F}) \tag{3.52}$$

In Eq. (3.51),

$$U_{\bar{\lambda}}(O\mathcal{F}) = \sqcup_{c \in C_1, C_2} \tilde{V}_c(O|\mathcal{F}) \tag{3.53}$$

$$U_{\bar{\lambda}}(\mathcal{F}) = \sqcup_{c \in C_1, C_2} \tilde{V}_c(\mathcal{F}) \tag{3.54}$$

Equations (3.53) and (3.54) are fuzzy likelihood and prior energy functions respectively. The fuzzy energy functions are based on join $\sqcup$ operation on fuzzy clique potentials over all possible cliques c. Also based on the following interval type fuzzy sets the energy function in above equations can be rewritten as the sum t-conorm operation on lower and upper membership functions in parallel.

$$\underline{h}(\mathcal{F}O) \propto \underline{h}(O\mathcal{F}) * \underline{h}(\mathcal{F}) \tag{3.55}$$

$$\bar{h}(\mathcal{F}O) \propto \bar{h}(O\mathcal{F}) * \bar{h}(\mathcal{F}) \tag{3.56}$$

The $*$ operator denotes the product t-norm in Eqs. (3.55) and (3.56). In principle clique potentials are selected arbitrarily if they decrease the values of the energy with an increase of matching degree [26]. To evaluate uncertain structural constraints in the labeling configuration, the fuzzy clique potentials are derived from T2 fuzzy Gaussian mixture models. From Eq. (3.57) and the Hammersley Clifford theorem [26], the relationship given in Eqs. (3.58) and (3.59) is obtained:

$$\underline{h}(O|\mathcal{F}) = \sqcap_{i=1}^{I} h_{\bar{\lambda}}(o_i, j) \tag{3.57}$$

$$h_{\bar{\lambda}}(o_i, j) = Z^{-1} e^{-\bar{V}_{c_1}(o_i|j, \lambda)} \tag{3.58}$$

$$h_{\bar{\lambda}}(o_i, o_{i\prime}, j, j\prime) = Z^{-1} e^{-\bar{V}_{c_2}(o_i, o_{i\prime}|j, j\prime, \lambda)} \tag{3.59}$$

In Eqs. (3.58) and (3.59) it is usually assumed that the normalizing constant $Z^{-1} = 1$. If $h_{\bar{\lambda}}(o_i, j)$ and $h_{\bar{\lambda}}(o_i, o_{i\prime}, j, j\prime)$ are T2 fuzzy Gaussian mixture models the following single pixel and pair pixel fuzzy likelihood clique potentials are derived from Eqs. (3.58) and (3.59).

$$\tilde{V}_{c_1}(o_i|j) = -\log \left[\sum_{i=1}^{M_s} w_{jm} N\left(o_i; \tilde{\mu}_{jm}, \Sigma_{jm}\right) \right] \tag{3.60}$$

$$\tilde{V}_{c_2}(o_i, o_{i\prime}|j, j\prime) = -\log \left[\sum_{i=1}^{M_s} w_{jj\prime m} N\left(o_{ii\prime}; \tilde{\mu}_{jj\prime m}, \Sigma_{jj\prime m}\right) \right] \tag{3.61}$$

In Eq. (3.61) the binary features $o_{ii'} = o_{i'} - o_i, i' \in \partial i$. The fuzzy prior clique potentials $\widetilde{V}_{c_1}(j)$ and $\widetilde{V}_{c_1}(j, j')$ encode the prior structural information in the labeling space. This prior structural information can be thought as a kind of smoothness [26] which encourages local labeling configurations consistent with the predefined structure at pixels and penalize those mismatched labeling configurations.

To find the single best labeling configuration $\mathcal{F}^*$ and the minimum fuzzy posterior energy $U_{\widetilde{\lambda}}(\mathcal{F}^*O)$ in Eqs. (3.52), (3.53) and (3.54) the T2 fuzzy relaxation labeling algorithm is achieved [26]. First the minimization of fuzzy posterior energy converted into the maximization of the following fuzzy gain function:

$$g_{\widetilde{\lambda}}(\mathcal{F}|O) = \bigsqcup_{j=1}^{J} \bigsqcup_{i=1}^{I} \left[\widetilde{K}_i(j) \sqcap f_i(j) + \bigsqcup_{j'=1}^{J} \max_{i' \in \partial i} \left[\widetilde{K}_{i,i'}(j, j') \sqcap f_i(j) \sqcap f_{i'}(j') \right] \right]$$

(3.62)

The fuzzy compatibility functions are defined by the fuzzy clique potentials as initialization part.

$$\widetilde{K}_i(j) = CONST_1 - \widetilde{V}_{c_1}(o_i|j) - \widetilde{V}_{c_1}(j)$$

(3.63)

$$\widetilde{K}_{i,i'}(j, j') = CONST_2 - \widetilde{V}_{c_2}(o_i, o_{i'}|j, j') - \widetilde{V}_{c_2}(j, j')$$

(3.64)

In Eq. (3.64) the constant $CONST_1$ and $CONST_2$ satisfy both fuzzy compatibility functions which are non-negative. The fuzzy gain function in above equation is the matching degree of the labeling configuration $\mathcal{F}$ to the observation O. In order to find the best $\mathcal{F}^*$ the maximizing labeling strengths in above equation are searched. The final labeling strength implies an assignment of the label to the observation [26]. The gradient of above equation updates the parameter $f_i^t(j)$ of T2 fuzzy relaxation labeling algorithm such that [26]:

$$\widetilde{q}_i(j) = \widetilde{K}_i(j) + \bigsqcup_{j'=1}^{J} \max_{i' \in \partial i} \left[\widetilde{K}_{i,i'}(j, j') \sqcap f_{i'}(j') \right]$$

(3.65)

The update in Eq. (3.65) is repeated until t reaches the fixed number T in T2 fuzzy relaxation labeling algorithm. As the compatibility functions contain both fuzzy likelihood and prior information, the solution of T2 fuzzy relaxation labeling algorithm does not depend on the initial labeling. Finally, the winner take all strategy is used to retrieve the best labeling configuration which ensures that each label j is assigned to merely one pixel i. The T2 fuzzy relaxation labeling algorithm is a context based algorithm because the labeling strength will increase only if its neighboring labeling strengths increase. According to the above equations, the meet $\sqcap$ and join $\sqcup$ operators can be rewritten by the t-norm and t-conorm operations on lower and upper membership functions in parallel. The product t-norm $*$ and sum t-conorm $\sum$ which involve the interval arithmetic are used here. For division between two intervals the center of the interval is divided unless otherwise stated. It is observed that the T2 fuzzy relaxation labeling algorithm has a polynomial complexity.

3.9 Other Soft Computing Techniques

In this section some other significant soft computing techniques for OCR systems are briefly presented. In late eighties there was a trend to integrate the technologies such as fuzzy sets, ANN, GA, rough sets etc. which would synergistically enhance the capability of each soft computing tool. This resulted in fusion and growth of various hybridization methods [9, 14, 20] such as neuro-fuzzy, neuro-genetic, neuro-fuzzy-genetic, rough-fuzzy-genetic, rough-neuro-fuzzy-genetic etc. Few such systems are already discussed in Sects. 3.5 and 3.6. Neuro-fuzzy hybridization is the most visible integration realized so far. The past few years have witnessed a rapid growth in number and variety of applications of fuzzy logic and ANN ranging from consumer electronics and industrial control to decision support systems and financial trading. Neuro-fuzzy modeling together with new driving force from stochastic, gradient-free optimization techniques such as GA and simulated annealing [20] forms constituents of soft computing which are aimed at solving real-world decision making, modeling and control problems. The integration of these technologies with rough sets has resulted in the birth and growth of hybrid systems which can handle uncertainty and vagueness stronger than either. These problems are usually imprecisely defined and require human intervention. Thus, neuro-fuzzy, rough sets and soft computing with their ability to incorporate human knowledge and to adapt their knowledge base via new optimization techniques play increasingly important roles in conception and design of hybrid intelligent systems. For further details on these techniques interested readers can refer [14, 18, 20].

References

1. Chaudhuri, A., Ghosh, S. K., Sentiment Analysis of Customer Reviews Using Robust Hierarchical Bidirectional Recurrent Neural Network, Book Chapter: Artificial Intelligence Perspectives in Intelligent Systems, Radek Silhavy, Roman Senkerik, Zuzana Kominkova Oplatkova, Petr Silhavy, Zdenka Prokopova, (Editors), Advances in Intelligent Systems and Computing, Springer International Publishing, Switzerland, Volume 464, pp 249–261, 2016.
2. Chaudhuri, A., Fuzzy Rough Support Vector Machine for Data Classification, International Journal of Fuzzy System Applications, 5(2), pp 26–53, 2016.
3. Chaudhuri, A., Modified Fuzzy Support Vector Machine for Credit Approval Classification, AI Communications, 27(2), pp 189–211, 2014.
4. Chaudhuri, A., De, Fuzzy Support Vector Machine for Bankruptcy Prediction, Applied Soft Computing, 11(2), pp 2472–2486, 2011.
5. Chaudhuri, A., Applications of Support Vector Machines in Engineering and Science, Technical Report, Birla Institute of Technology Mesra, Patna Campus, India, 2011.
6. Chaudhuri, A., Some Experiments on Optical Character Recognition Systems for different Languages using Soft Computing Techniques, Technical Report, Birla Institute of Technology Mesra, Patna Campus, India, 2010.
7. Chaudhuri, A., De, K., Job Scheduling using Rough Fuzzy Multi-Layer Perception Networks, Journal of Artificial Intelligence: Theory and Applications, 1(1), pp 4–19, 2010.

8. Chaudhuri, A., De, K., Chatterjee, D., Discovering Stock Price Prediction Rules of Bombay Stock Exchange using Rough Fuzzy Multi-Layer Perception Networks, Book Chapter: Forecasting Financial Markets in India, Rudra P. Pradhan, Indian Institute of Technology Kharagpur, (Editor), Allied Publishers, India, pp 69–96, 2009.
9. Chaudhuri, A., Studies in Applications of Soft Computing to some Optimization Problems, PhD Thesis, Netaji Subhas Open University, Kolkata, India, 2010.
10. Cheriet, M., Kharma, N., Liu, C. L., Suen, C. Y., Character Recognition Systems: A Guide for Students and Practitioners, John Wiley and Sons, 2007.
11. De, R. K., Pal, N. R., Pal, S. K. Feature Analysis: Neural Network and Fuzzy Set Theoretic Approaches, Pattern Recognition, 30(10), pp 1579–1590, 1997.
12. Goldberg, D. E., Genetic Algorithms in Search, Optimization and Machine Learning, Reading, Mass, Addison Wesley, 1989.
13. Haykin, S., Neural Networks and Learning Machines, 3rd Edition, Prentice Hall, 2008.
14. Jang, J. S. R., Sun, C. T., Mizutani, E., Neuro-Fuzzy and Soft Computing: A Computational Approach to Learning and Machine Intelligence, Prentice Hall, 1997.
15. Klir, G. J., Yuan, B., Fuzzy Sets and Fuzzy Logic, Prentice Hall, New Jersey, 1995.
16. Mitchell, M., An Introduction to Genetic Algorithms, MIT Press, 1998.
17. Pal, S. K., Mitra, S., Mitra, P., Rough-Fuzzy Multilayer Perception: Modular Evolution, Rule Generation and Evaluation, IEEE Transactions on Knowledge and Data Engineering, 15(1), pp 14–25, 2003.
18. Pal, S. K., Soft Computing Pattern Recognition: Principles, Integrations and Data Mining, In T. Tassano et al. (Editors), New Frontiers in Artificial Intelligence, Lecture Notes in Computer Science, Springer Verlag, Berlin, LNCS 2253, pp 261–271, 2001.
19. Polkowski, L, Rough Sets – Mathematical Foundations, Advances in Intelligent and Soft Computing, Springer Verlag, 2002.
20. Pratihar, D. K., Soft Computing, Alpha Science International Limited, 2007.
21. Yen, J., Langari, R., Fuzzy Logic: Intelligence, Control and Information, Pearson Education, 2005.
22. Young, T. Y., Fu, K. S., Handbook of Pattern Recognition and Image Processing, Academic Press, 1986.
23. Zadeh, L. A., Fuzzy Logic, Neural Networks and Soft Computing, Communications of the ACM, 37(3), pp 77–84, 1994.
24. Zadeh, L. A., Fuzzy Sets as a Basis for a Theory of Possibility, Fuzzy Sets and Systems, 1(1), pp 3–28, 1978.
25. Zadeh, L. A., Fuzzy Sets, Information and Control, 8(3), pp 338–353, 1965.
26. Zeng, J., Liu, Z. Q., Type-2 Fuzzy Markov Random Fields and their Application to Handwritten Chinese Character Recognition, IEEE Transactions on Fuzzy Systems, 16(3), pp 747–760, 2008.
27. Zimmermann, H. J., Fuzzy Set Theory and its Applications, 4th Edition, Kluwer Academic Publishers, Boston, 2001.

Chapter 4
Optical Character Recognition Systems for English Language

Abstract The optical character recognition (OCR) systems for English language were the most primitive ones and occupy a significant place in pattern recognition. The English language OCR systems have been used successfully in a wide array of commercial applications. The different challenges involved in the OCR systems for English language is investigated in this chapter. The pre-processing activities such as binarization, noise removal, skew detection and correction, character segmentation and thinning are performed on the datasets considered. The feature extraction is performed through discrete cosine transformation. The feature based classification is performed through important soft computing techniques viz fuzzy multilayer perceptron (FMLP), rough fuzzy multilayer perceptron (RFMLP), fuzzy support vector machine (FSVM) and fuzzy rough support vector machine (FRSVM). The superiority of soft computing techniques is demonstrated through the experimental results.

Keywords English language OCR · FMLP · RFMLP · FSVM · FRSVM

4.1 Introduction

The first and the most primitive optical character recognition (OCR) systems were developed for English language [1, 2]. English being the most widely spoken language across the globe [3], OCR systems for English language occupy a significant place in pattern recognition [4]. English language OCR systems have been used successfully in a wide array of commercial applications [5]. The character recognition of English language has a great potential in data and word processing. Some commonly used applications are automated postal address and ZIP code reading, data acquisition in bank checks, processing of archived institutional records etc. The American standardization of OCR character set for English language was provided through OCR-A as shown in Fig. 4.1. The European standardization was provided by OCR-B character set as shown in Fig. 4.2. In the recent

A. Chaudhuri et al., *Optical Character Recognition Systems for Different Languages with Soft Computing*, Studies in Fuzziness and Soft Computing 352, DOI 10.1007/978-3-319-50252-6_4

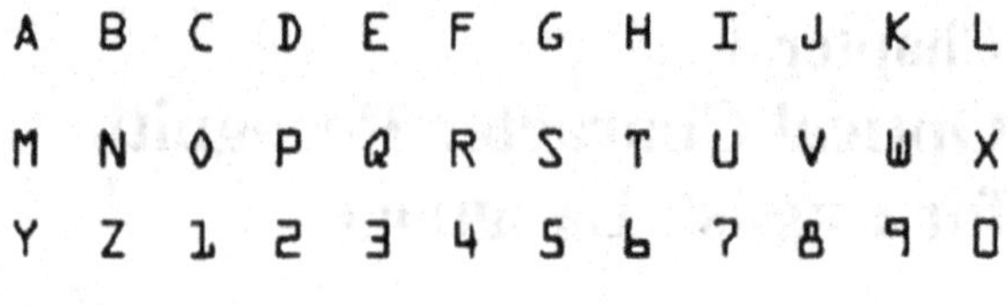

Fig. 4.1 OCR–A font

A B C D E F G H I J K L

M N O P Q R S T U V W X

Y Z 1 2 3 4 5 6 7 8 9 0

Fig. 4.2 OCR–B font

A B C D E F G H I J K L

M N O P Q R S T U V W X

Y Z 1 2 3 4 5 6 7 8 9 0

years, OCR for English language has gained a considerable momentum, as the need for converting the scanned images into computer recognizable formats such as text documents has variety of applications. English language based OCR systems is thus one of the most fascinating and challenging areas of pattern recognition with various practical applications [1].

The OCR process for any language involves extraction of defined characteristics called features to classify an unknown character into one of the known classes [1, 5]. As such any good OCR system is best defined in terms of the success of feature extraction and classification tasks. The same is true for English language. The process becomes tough in case the language has dependencies where some characters look identical. Thus the classification becomes a big challenge.

In this chapter we start the investigation of OCR systems considering the different challenges involved in English language. The different pre-processing activities such as binarization, noise removal, skew detection and correction, character segmentation and thinning are performed on the considered datasets [6]. The feature extraction is performed through discrete cosine transformation. The feature based classification is performed through important soft computing techniques viz fuzzy multilayer perceptron (FMLP) [1], rough fuzzy multilayer perceptron (RFMLP) [1, 7, 8, 9] and two support vector machine (SVM) based methods such as fuzzy support vector machine (FSVM) [10, 11] and fuzzy rough support vector machine (FRSVM) [12, 13]. The experimental results demonstrate the superiority of soft computing techniques over the traditional methods.

This chapter is structured as follows. In Sect. 4.2 a brief discussion about the English language script and datasets used for experiments are presented. The different challenges of OCR for English language are highlighted in Sect. 4.3. The next section illustrates the data acquisition. In Sect. 4.5 different pre-processing activities on the datasets such as binarization, noise removal, skew detection and correction, character segmentation and thinning are presented. This is followed by a discussion of feature extraction on English language dataset in Sect. 4.6. The Sect. 4.7 explains the state of art of OCR for English language in terms of feature based classification methods. The corresponding experimental results are given in Sect. 4.8. Finally Sect. 4.9 concludes the chapter with some discussions and future research directions.

4.2 English Language Script and Experimental Dataset

In this section we present brief information about the English language script and the dataset used for experiments. English is West Germanic language [3] that was first spoken in early medieval England and is now a global lingua franca. It is the official language of almost 60 sovereign states. It is most commonly spoken language in United Kingdom, United States, Canada, Australia, Ireland, and New Zealand and widely spoken language in countries in Caribbean, Africa, and South Asia. It is the third most common native language in the world. It is most widely learned second language and is the official language of United Nations, European Union and many other international organisations. English has developed over the course of more than 1,400 years. The earliest form of English which is a set of Anglo-Frisian dialects brought to Great Britain by Anglo-Saxon settlers in 5th century is the old English. The middle English began in late 11th century with the Norman conquest of England. The early modern English began in late 15th century with the introduction of the printing press to London and the King James Bible as well as the Great Vowel Shift. Through the worldwide influence of British Empire, modern English spread around the world from 17th to mid-20th centuries. Through all types of printed and electronic media as well as the emergence of United States as global superpower, English has become the leading language of international discourse and the lingua franca in many regions and in several professional contexts.

The English language dataset used for performing OCR experiments is the IAM dataset and is adapted from [6]. IAM database contains English text which is used here to for training and testing. The database was first used for experiments resulting in a publication at ICDAR 1999. Thereafter the database was regularly used for various experiments. The database contains unconstrained handwritten text which are scanned at a resolution of 300 dpi and saved as PNG images with 256 gray levels. The Fig. 4.3 shows a sample from the database. IAM database is structured as follows:

(a) 657 writers contributed samples of their handwriting
(b) 1539 pages of scanned text
(c) 5685 isolated and labeled sentences
(d) 13,353 isolated and labeled text lines
(e) 15,320 isolated and labeled words

Further details are available at [6].

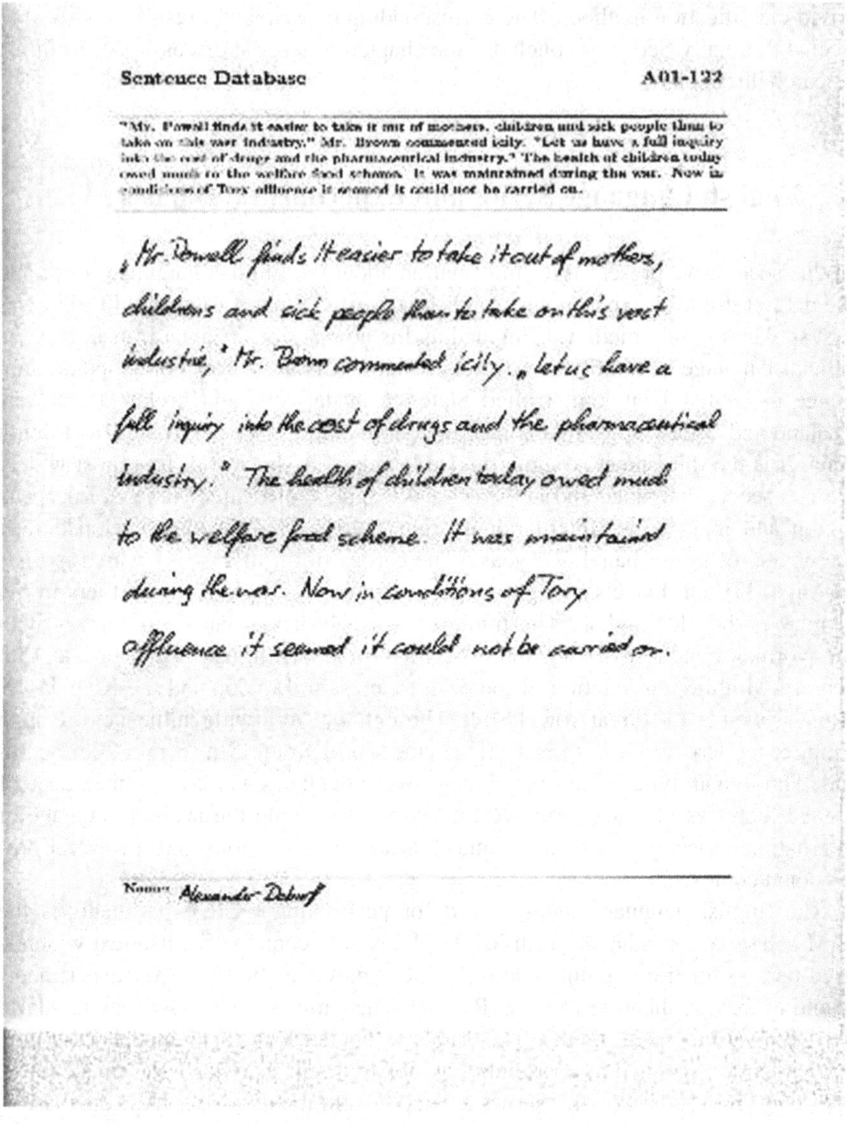

Sentence Database A01-122

"Mr. Powell finds it easier to take it out of mothers, children and sick people than to take on this vast industry." Mr. Brown commented icily. "Let us have a full inquiry into the cost of drugs and the pharmaceutical industry." The health of children today owed much to the welfare food scheme. It was maintained during the war. Now in conditions of Tory affluence it seemed it could not be carried on.

,,Mr. Powell finds it easier to take it out of mothers, childrens and sick people than to take on this vast industrie," Mr. Brown commented icilly. ,,letus have a full inquiry into the cost of drugs and the pharmaceutical industry." The health of children today owed much to the welfare food scheme. It was maintained during the war. Now in conditions of Tory affluence it seemed it could not be carried on.

Name: Alexander Dohrf

Fig. 4.3 A sample text from IAM database

4.3 Challenges of Optical Character Recognition Systems for English Language

OCR for English language has become one of the most successful applications of technology in pattern recognition and artificial intelligence. OCR for English language has been the topic of intensive research for more than five decades [1, 4, 5]. The most primitive OCR systems were developed in English language. Considering the important aspects of versatility, robustness and efficiency, commercial OCR systems are generally divided into four generations [1] as highlighted in Chap 2. It is to be noted that this categorization refers specifically to OCRs of English language.

Despite decades of research and existence of established commercial OCR products based on English language, the output from such OCR processes often contains errors. The more highly degraded is input, the greater is error rate. Since inputs form the first stage in a pipeline where later stages are designed to support sophisticated information extraction and exploitation applications, it is important to understand the effects of recognition errors on downstream analysis routines. Few questions are required to be addressed in this direction. They are as follows:

(a) Are all recognition errors equal in impact or some are worse than others?
(b) Can the performance of each stage be optimized in isolation or the end-to-end system should be considered?
(c) In balancing the trade-off between the risk of over and under segmenting characters during OCR where should the line be drawn to maximize overall performance?

The answers to these questions often influence the way OCR systems are designed and build for analysis [5].

The English language OCR system converts numerous published books in English language into editable computer text files. The latest research in this area has grown to incorporate some new methodologies to overcome the complexity of English writing style. All these algorithms have still not been tested for complete characters of English alphabet. Hence, there is a quest for developing an OCR system which handles all classes of English text and identify characters among these classes increasing versatility, robustness and efficiency in commercial OCR systems. The recognition of printed English characters is itself a challenging problem since there is a variation of the same character due to change of fonts or introduction of different types of noises. There may be noise pixels that are introduced due to scanning of the image. A significant amount of research has been done towards text data processing in English language from noisy sources [4]. The majority of the work has focused predominately on errors that arise during speech recognition systems [5]. Several research papers have appeared which examines the noise problem from variety of perspectives with most emphasizing issues that are inherent in written and spoken English language [1]. However, there has been less work concentrating on noise induced by OCR. Some earlier works by [14] show that moderate error rates have little impact on effectiveness of traditional information retrieval measures. However, this conclusion is tied to certain assumptions about information retrieval through bag of words, OCR error rate which may not be too high and length of documents which may not be too short. Some other notable research works in this direction are given in [15, 16]. All these works try to address some significant issues involved in OCR systems for English language such as error correction, performance evaluation etc. involving flexible and rigorous mathematical treatment [1]. Besides this any English character can be represented in variety of fonts and sizes as per the needs and requirements of application. Further the character with same font and size may also be bold face character as well as normal one [5]. Thus the width of stroke also significantly affects recognition process. Therefore, a good character recognition approach for English language [1, 17, 18, 19]:

(a) Must eliminate noise after reading binary image data
(b) Smooth image for better recognition
(c) Extract features efficiently
(d) Train the system and
(e) Classify patterns accordingly.

4.4 Data Acquisition

The progress in automatic character recognition systems in English language is generally evolved in two categories according to the mode of data acquisition which can be either online or offline character recognition systems. Offline character recognition captures data from paper through optical scanners or cameras whereas online recognition systems utilize digitizers which directly capture writing with the order of strokes, speed, pen-up and pen-down information. As such the scope of this text is restricted to OCR systems, we confine our discussion to offline character recognition [1] for English language. The English language datasets used in this research is mentioned in Sect. 4.2.

4.5 Data Pre-processing

Once the data has been acquired properly we proceed to pre-process the data. In pre-processing stage [1] a series of operations are performed viz binarization, noise removal, skew detection, character segmentation and thinning or skeletonization. The main objective of pre-processing is to organize information so that the subsequent character recognition task becomes simpler. It essentially enhances the image rendering it suitable for segmentation.

4.5.1 Binarization

Binarization [5] is an important first step in character recognition process. A large number of binarization techniques are available in the literature [4] each of which is appropriate to particular image types. Its goal is to reduce the amount of information present in the image and keep only the relevant information. Generally the binarization techniques of gray scale images are classified into two categories viz overall threshold where single threshold is used in the entire image to form two classes (text and background) and local threshold where values of thresholds are determined locally (pixel-by-pixel or region-by-region). Here, we calculate threshold Th of each pixel locally by [1]:

$$Th = (1 - k) \cdot m + k \cdot m + k \cdot \frac{\sigma}{R(m - M)} \tag{4.1}$$

In Eq. (4.1) k is set to 0.5; m is the average of all pixels in window; M is the minimum image grey level; σ is the standard deviation of all pixels in window and R is the maximum deviation of grayscale on all windows.

4.5.2 Noise Removal

The scanned text documents often contain noise that arises due to printer, scanner, print quality, document age etc. Therefore, it is necessary to filter noise [4] before the image is processed. Here a low-pass filter is used to process the image [1] which is used for later processing. The main objective in the design of a noise filter is that it should remove as much noise as possible while retaining the entire signal [5].

4.5.3 Skew Detection and Correction

When a text document is fed into scanner either mechanically or manually a few degrees of tilt or skew is unavoidable. In skew angle the text lines in digital image make angle with horizontal direction. A number of methods are available in literature for identifying image skew angles [4]. They are basically categorized on the basis of projection profile analysis, nearest neighbor clustering, Hough transform, cross correlation and morphological transforms. The aforementioned methods correct the detected skew. Some of the notable research works in this direction are available in [1]. Here Hough transform is used for skew detection and correction [1].

4.5.4 Character Segmentation

Once the text image is binarized, noise removed and skew corrected, the actual text content is extracted. This process leads to character segmentation [17, 18, 19]. The commonly used segmentation algorithms in this direction are connected

Fig. 4.4 An image before
and after thinning

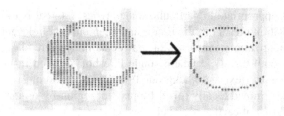

component labeling, x-y tree decomposition, run length smearing and Hough transform [4]. Here Hough transform is used for character segmentation [1].

4.5.5 Thinning

The character segmentation process is followed by thinning or skeletonization. In thinning one-pixel-width representation or skeleton of an object is obtained by preserving connectedness of the object and its end points [20]. The thinning process reduces image components to their essential information so that further analysis and recognition are facilitated. For instance, an alphabet can be hand-written with different pens giving different stroke thicknesses but information presented is same. This enables easier subsequent detection of pertinent features. As an illustration consider letter *e* shown in Fig. 4.4 before and after thinning. A number of thinning algorithms have been used in the past with considerable success. The most common algorithm used is the classical hilditch algorithm and its variants [1]. Here hilditch algorithm is used for thinning [5]. For recognizing large graphical objects with filled regions which are often found in logos, region boundary detection is useful but for small regions corresponding to individual characters neither thinning nor boundary detection is performed. Rather entire pixel array representing the region is forwarded to subsequent stage of analysis.

4.6 Feature Extraction

The heart of any OCR system is the formation of feature vector which is used in recognition stage. The same is true for English language OCR system. This phase is designed to extract the features from segmented areas of image containing characters to be recognized, especially traits that serve to distinguish an area corresponding to a letter from an area corresponding to other letters. The feature extraction can thus be considered as finding a set of parameters or features that define the shape of underlying character as precisely and uniquely as possible. The term feature extraction is often replaced by feature selection which refers to algorithms that select the best subset of input feature set. These methods create new features based on transformations or combination of original features [21, 23]. The features are selected in such a way so that they help in character discrimination. The selection of feature extraction methods is probably an important factor in achieving high recognition performance. A large number of OCR based feature extraction methods are available in literature [4, 5]. However, the selected methods depend on the given application. There is no universally accepted set of feature vectors in OCR. The features that capture topological and geometrical shape information are the most desired ones. The features that capture spatial distribution of black (text) pixels are also very important [4]. The two most important types of feature extraction approaches which has been successfully used in the past towards OCR of English language are structural and statistical techniques [13].

For extracting the features that define characters in the image, discrete cosine transformation [20] is used here. The discrete cosine transformation converts a signal into its elementary frequency components. Each line of M pixels from character image is represented as a sum of M weighted cosine functions assessed in discrete points as:

$$T_i = \sum_{j=0}^{M-1} \frac{\sqrt{2}C_j}{\sqrt{M}} s_j \cos \frac{(2i+1)j\pi}{2M} \qquad \forall \quad 0 \leq i \leq M \qquad (4.2)$$

In Eq. (4.2) $C_j = \frac{1}{\sqrt{2}}$ for $j = 0$ otherwise $C_j = 1$. The above equation is true for one dimensional case. In two dimensional case, we consider a matrix S of 8×8 elements defined by [20]:

$$T_{i,k} = \frac{1}{4} C_i C_k \sum_{p=0}^{7} \sum_{q=0}^{7} s_{p,q} \cos \frac{(2p+1)k\pi}{16} \cos \frac{(2q+1)i\pi}{16} \qquad (4.3)$$

In Eq. (4.3) $C_i = C_k = \frac{1}{\sqrt{2}}$ for $i,j = 0$ otherwise $C_i = C_k = 1$. The transformed matrix elements with lower indices correspond to coarser details and those with higher indices to finer details in image. On analyzing matrix T obtained by processing different blocks of an image, it is observed that in upper left corner of matrix there are high values (positive or negative) and in bottom right corner the values start to decline and tend towards 0. Next the certain elements in the array are selected. The first operation is performed to order the elements of matrix into a one dimensional array to highlight as many values of zero as possible. The ordering is done by reading the matrix in zigzag. To extract necessary features for character recognition the first N values from this array are selected. As N increases so does the recognition accuracy, but that happens at the expense of increasing the training time of classifiers [1].

The first n components of discrete cosine transformation of a segmented area of image are considered to describe the features of image. Here, certain statistical details of the area are added to discrete cosine transformation components to define its features specified in terms of (a) number of black pixels from a matrix (also called on pixels) (b) mean of horizontal positions of all the on pixels relative to the centre of image and its width (c) mean of vertical positions of all the on pixels relative to the centre of image and its height (d) mean of horizontal distances between on pixels (e) mean of vertical distances between on pixels (f) mean product between vertical and horizontal distances of on pixels (g) mean product between the square of horizontal and vertical distances between all on pixels (h) mean product between the square of vertical and horizontal distances between all on pixels (i) mean number of margins met by scanning the image from left to right (j) sum of vertical positions of margins met by scanning the image from left to right (k) mean number of margins met by scanning the image from bottom to top (l) sum of horizontal positions of margins met by scanning the image from top to bottom. One important operation implemented by this phase is the normalization of results obtained till now so as they correspond to the format accepted by the feature based classification methods presented in Sect. 4.7. The feature extraction techniques are evaluated here based on [1]:

(a) Robustness against noise, distortion, size and style or font variation
(b) Speed of recognition
(c) Complexity of implementation
(d) The independence of feature set is without any supplementary techniques.

4.7 Feature Based Classification: Sate of Art

After extracting the essential features from the pre-processed character image we concentrate on the feature based classification methods for OCR of English language. Keeping in view of the wide array of feature based classification methods for OCR, these methods are generally grouped into following four broad categories [15, 22, 23]:

(a) Statistical methods
(b) ANN based methods
(c) Kernel based methods
(d) Multiple classifier combination methods.

We discuss here the work done on English characters using some of the above mentioned classification methods. All the classification methods used here are soft computing based techniques. The next three subsections highlight the feature based classification based on FMLP, RFMLP [7, 8], FSVM [10, 11] and FRSVM [12], techniques. These methods are already discussed in Chap 3. For further details interested readers can refer [1, 13].

4.7.1 Feature Based Classification Through Fuzzy Multilayer Perceptron

The English language characters are recognized here through FMLP [11] which is discussed in Chap 3. In recent past FMLP has been used with considerable success [9]. We use diagonal feature extraction scheme for drawing of features from the characters. For classification stage we use these features extracted. FMLP used as a feed forward network with back propagation ANN having two hidden layers. The architecture used is 70–100–100–52 [1] for classification. The two hidden layers and output layer uses tangent sigmoid as the activation function. The feature vector is denoted by $F = (f_1, \ldots, f_m)$ where m denotes number of training images and each f has a length of 70 which represent the number of input nodes. The 52 neurons in the output layer correspond to the 26 English alphabets both uppercase and lowercase. The network training parameters are briefly summarized as [1]:

(a) Input nodes: 70
(b) Hidden nodes: 100 at each layer
(c) Output nodes: 52
(d) Training algorithm: Scaled conjugate gradient backpropagation
(e) Performance function: Mean Square Error
(f) Training goal achieved: 0.000004
(g) Training epochs: 4000.

4.7.2 Feature Based Classification Through Rough Fuzzy Multilayer Perceptron

In similar lines to the English language character recognition through FMLP in Sect. 4.7.1, RFMLP [7–9] discussed in Chap 3 is used here to recognize the English language characters. The diagonal feature extraction scheme is used for drawing of features from the characters. For classification stage we use these features extracted. RFMLP used here is a feed forward network with back propagation ANN having four hidden layers. The architecture used is 70–100–100–100–100–100–52 [1] for classification. The four hidden layers and output layer uses tangent sigmoid as the activation function. The feature vector is again denoted by $F = (f_1, \ldots, f_m)$ where m denotes number of training images and each f has a length of 70 which represent the number of input nodes. The 52 neurons in the output layer correspond to the 26 English alphabets both uppercase and lowercase. The network training parameters are briefly summarized as [1]:

(a) Input nodes: 70
(b) Hidden nodes: 100 at each layer
(c) Output nodes: 52
(d) Training algorithm: Scaled conjugate gradient backpropagation

(e) Performance function: Mean Square Error
(f) Training goal achieved: 0.000002
(g) Training epochs: 4000.

4.7.3 Feature Based Classification Through Fuzzy and Fuzzy Rough Support Vector Machines

SVM based methods have shown a considerable success in feature based classification [13]. With this motivation we use here FSVM [10, 11] and FRSVM [12] for feature classification task. Both FSVM and FRSVM offer the possibility of selecting different types of kernel functions such as sigmoid, RBF, linear functions and determining the best possible values for these kernel parameters [1]. After selecting the kernel type and its parameters, FSVM and FRSVM are trained with the set of features obtained from other phases. Once the training gets over, FSVM and FRSVM are used to classify new character sets. The implementation is achieved through LibSVM library [24].

4.8 Experimental Results

In this section, the experimental results for soft computing tools viz FMLP, RFMLP, FSVM and FRSVM on the English language dataset [6] highlighted in Sect. 4.2 are presented. The prima face is to select the best OCR system for English language [1].

4.8.1 Fuzzy Multilayer Perceptron

The experimental results for the English characters both uppercase and lowercase using FMLP are shown in the Table 4.1. Here, the English characters are categorized as ascenders (characters such as b, d, h, k, J, L etc.) and descenders (characters such as g, p, q, j, t, y, A, F, P etc.). FMLP shows better results than the traditional methods [1]. The testing algorithm is applied to all standard cases of characters [6] in various samples. An accuracy of around 98% has been achieved in all the cases. However, after successful training and testing of algorithm the following flaws are encountered [1]:

(a) There may an instance when there is an occurrence of large and disproportionate symmetry in both ascenders as well as descenders as shown in Fig. 4.5a, b.
(b) There may be certain slants in certain characters which results in incorrect detection of characters as shown in Fig. 4.6.

Table 4.1 The experimental results for the English characters using FMLP

Characters	Successful recognition (%)	Unsuccessful recognition (%)	No recognition (%)
a	98	1	1
b	97	1	2
c	99	1	0
d	97	1	2
e	98	1	1
f	99	1	0
g	97	1	2
h	98	1	1
i	99	1	0
j	99	1	0
k	98	1	1
l	99	1	0
m	98	1	1
n	98	1	1
o	98	1	1
p	97	1	2
q	97	1	2
r	99	1	0
s	98	1	1
t	99	1	0
u	98	1	1
v	98	1	1
w	97	1	2
x	98	1	1
y	99	1	0
z	99	1	0
A	98	1	1
B	98	1	1
C	99	1	0
D	98	1	1
E	98	1	1
F	98	1	1
G	98	1	1
H	98	1	1
I	99	1	1
J	99	1	0
K	98	1	1
L	99	1	0
M	98	1	1
N	98	1	1

(continued)

Table 4.1 (continued)

Characters	Successful recognition (%)	Unsuccessful recognition (%)	No recognition (%)
O	98	1	1
P	98	1	1
Q	98	1	1
R	98	1	1
S	98	1	1
T	99	1	0
U	99	1	0
V	98	1	1
W	98	1	1
X	98	1	1
Y	99	1	0
Z	98	1	1

Fig. 4.5 The disproportionate symmetry in characters 'p' and 't'

(a)

(b)

Fig. 4.6 The character 't' with slant

Fig. 4.7 The uneven pixel combination for character 'k' (region-i on the *left image* has less white space than the *right image*)

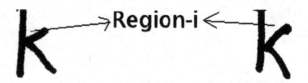

(c) Sometimes one side of the image may have less white space and more pixel concentration whereas other side may have more white space and less pixel concentration. Due to this some characters are wrongly detected as shown in Fig. 4.7.

Table 4.2 The experimental results for the English characters using RFMLP

Characters	Successful recognition (%)	Unsuccessful recognition (%)	No recognition (%)
a	99	1	0
b	99	1	0
c	99	1	0
d	99	1	0
e	99	1	0
f	99	1	0
g	98	1	1
h	99	1	0
i	99	1	0
j	99	1	0
k	99	1	0
l	99	1	0
m	98	1	1
n	98	1	1
o	99	1	0
p	99	1	0
q	99	1	0
r	99	1	0
s	99	1	0
t	99	1	0
u	99	1	0
v	99	1	0
w	98	1	1
x	98	1	1
y	99	1	0
z	99	1	0
A	99	1	0
B	99	1	0
C	99	1	0
D	99	1	0
E	99	1	0
F	99	1	0
G	99	1	0
H	99	1	0
I	99	1	0
J	99	1	0
K	99	1	0
L	99	1	0
M	98	1	1
N	98	1	1

(continued)

Table 4.2 (continued)

Characters	Successful recognition (%)	Unsuccessful recognition (%)	No recognition (%)
O	99	1	0
P	99	1	0
Q	99	1	0
R	99	1	0
S	99	1	0
T	99	1	0
U	99	1	0
V	99	1	0
W	98	1	1
X	98	1	1
Y	99	1	0
Z	99	1	0

4.8.2 Rough Fuzzy Multilayer Perceptron

All the aforementioned flaws in FMLP are taken care of through RFMLP [7–9] where rough and fuzzy sets combination actively helps in improving the overall results [1]. The experimental results for the English characters both uppercase and lowercase using RFMLP are shown in the Table 4.2. An accuracy of around 99% has been achieved in all the cases.

4.8.3 Fuzzy and Fuzzy Rough Support Vector Machines

FSVM [10, 11] and FRSVM [12] show promising results for feature classification task through different types of kernel functions by selecting the best possible values for kernel parameters [13]. For testing accuracy of the system in the first test case scenario we use an image which contained 100 small letters as shown in Fig. 4.8. The training set is constructed through two images containing 40 examples of each small letter in the English alphabet which took around 17.69 s. The results are presented in Tables 4.3 and 4.4. The parameter C is regularization parameter, ρ is bias term, κ is kernel function, σ is smoothing function which reduces variance and ϕ is mapping function in feature space. Further discussion on these parameters is available in [1].

For the next test case scenario we use for training only the features corresponding to capital letters. The image used for testing contained 100 letters as shown in Fig. 4.9. The training set is constructed through two images containing 40 examples of each capital letter in the English alphabet which took around 18.79 s. The results are presented in Tables 4.5 and 4.6.

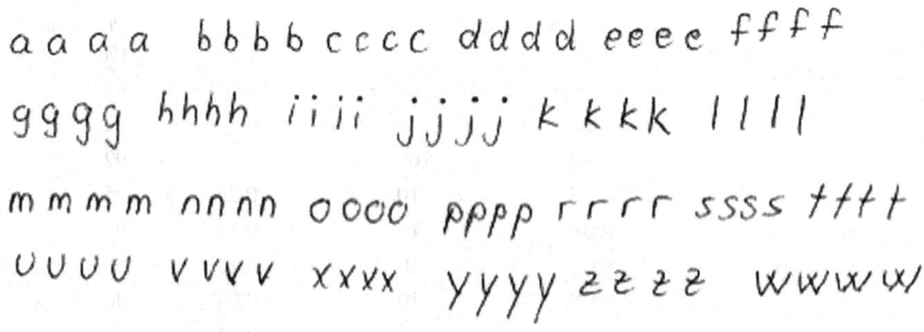

Fig. 4.8 The test image for small letters

Table 4.3 The training set results corresponding to small letters (for FSVM)

Kernel function	C	ρ	κ	σ	ϕ	Precision (%)
Linear	1	–	–	–	–	92
Linear	10	–	–	–	–	90
Linear	100	–	–	–	–	91
RBF	10	–	–	0.25	–	94
RBF	10	–	–	0.15	–	94
RBF	10	–	–	0.10	–	95
RBF	10	–	–	0.05	–	95
RBF	10	–	–	0.03	–	96
RBF	10	–	–	0.02	–	96
RBF	10	–	–	0.01	–	97
RBF	10	–	–	0.005	–	97
Polynomial	10	2	2	–	–	96
Polynomial	10	2	4	–	–	96
Polynomial	10	2	1	–	–	96
Polynomial	10	2	0.5	–	–	94
Polynomial	10	3	2	–	–	90
Polynomial	10	3	4	–	–	91
Polynomial	10	3	1	–	–	94
Polynomial	10	3	0.5	–	–	94
Polynomial	10	4	2	–	–	94
Polynomial	10	4	4	–	–	93
Polynomial	10	4	1	–	–	94
Polynomial	10	4	0.5	–	–	94
Sigmoid	10	–	0.5	–	1	92
Sigmoid	10	–	0.5	–	5	93
Sigmoid	10	–	0.2	–	1	94
Sigmoid	10	–	0.7	–	1	95

Table 4.4 The training set results corresponding to small letters (for FRSVM)

Kernel function	C	ρ	κ	σ	ϕ	Precision (%)
Linear	1	–	–	–	–	93
Linear	10	–	–	–	–	92
Linear	100	–	–	–	–	92
RBF	10	–	–	0.25	–	95
RBF	10	–	–	0.15	–	95
RBF	10	–	–	0.10	–	96
RBF	10	–	–	0.05	–	96
RBF	10	–	–	0.03	–	97
RBF	10	–	–	0.02	–	97
RBF	10	–	–	0.01	–	98
RBF	10	–	–	0.005	–	98
Polynomial	10	2	2	–	–	98
Polynomial	10	2	4	–	–	98
Polynomial	10	2	1	–	–	97
Polynomial	10	2	0.5	–	–	95
Polynomial	10	3	2	–	–	92
Polynomial	10	3	4	–	–	93
Polynomial	10	3	1	–	–	95
Polynomial	10	3	0.5	–	–	95
Polynomial	10	4	2	–	–	95
Polynomial	10	4	4	–	–	94
Polynomial	10	4	1	–	–	95
Polynomial	10	4	0.5	–	–	95
Sigmoid	10	–	0.5	–	1	93
Sigmoid	10	–	0.5	–	5	94
Sigmoid	10	–	0.2	–	1	95
Sigmoid	10	–	0.7	–	1	96

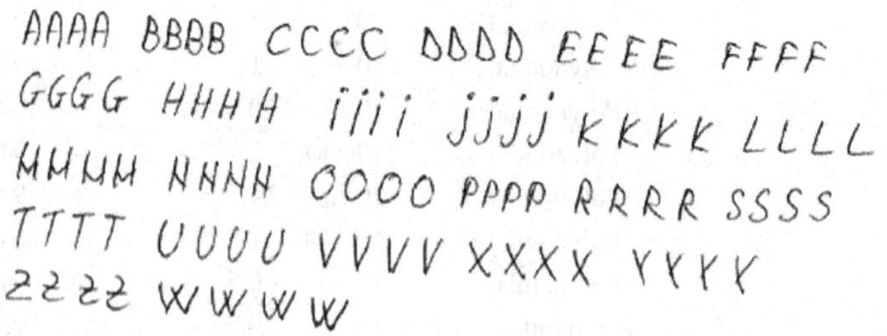

Fig. 4.9 The test image for capital letters

Table 4.5 The training set
results corresponding to
capital letters (for FSVM)

Kernel function	C	ρ	κ	σ	ϕ	Precision (%)
Linear	1	–	–	–	–	93
Linear	10	–	–	–	–	94
Linear	100	–	–	–	–	92
RBF	10	–	–	0.25	–	96
RBF	10	–	–	0.15	–	95
RBF	10	–	–	0.10	–	97
RBF	10	–	–	0.05	–	94
RBF	10	–	–	0.03	–	92
RBF	10	–	–	0.02	–	92
RBF	10	–	–	0.01	–	91
RBF	10	–	–	0.005	–	94
Polynomial	10	2	2	–	–	89
Polynomial	10	2	4	–	–	94
Polynomial	10	2	1	–	–	92
Polynomial	10	2	0.5	–	–	94
Polynomial	10	3	2	–	–	94
Polynomial	10	3	4	–	–	89
Polynomial	10	3	1	–	–	96
Polynomial	10	3	0.5	–	–	92
Polynomial	10	4	2	–	–	95
Polynomial	10	4	4	–	–	94
Polynomial	10	4	1	–	–	92
Polynomial	10	4	0.5	–	–	92
Sigmoid	10	–	0.5	–	1	89
Sigmoid	10	–	0.2	–	1	89

For the final test case scenario we use for training the features corresponding to both small and capital letters. The images used for testing are the ones used in the first and second test cases. The training set construction took around 36.96 s. The results are presented in Tables 4.7 and 4.8.

A comparative performance of the soft computing techniques (FMLP, RFMLP, FSVM, FRSVM) used for English language with the traditional techniques (MLP, SVM) is provided in Fig. 4.10 for samples of 30 datasets. It is to be noted that the architecture of MLP used for classification is 70–100–100–52 [1] and sigmoid kernel is used with SVM.

All the tests are conducted on PC having Intel P4 processor with 4.43 GHz, 512 MB DDR RAM @ 400 MHz with 512 KB cache. The training set construction was the longest operation of the system where the processor was loaded to 25% and the application occupied around 54.16 MB of memory. During idle mode the application consumes 43.02 MB of memory.

Table 4.6 The training set results corresponding to capital letters (for FRSVM)

Kernel function	C	ρ	κ	σ	ϕ	Precision (%)
Linear	1	–	–	–	–	94
Linear	10	–	–	–	–	95
Linear	100	–	–	–	–	92
RBF	10	–	–	0.25	–	97
RBF	10	–	–	0.15	–	96
RBF	10	–	–	0.10	–	98
RBF	10	–	–	0.05	–	95
RBF	10	–	–	0.03	–	93
RBF	10	–	–	0.02	–	92
RBF	10	–	–	0.01	–	92
RBF	10	–	–	0.005	–	95
Polynomial	10	2	2	–	–	92
Polynomial	10	2	4	–	–	95
Polynomial	10	2	1	–	–	93
Polynomial	10	2	0.5	–	–	95
Polynomial	10	3	2	–	–	96
Polynomial	10	3	4	–	–	91
Polynomial	10	3	1	–	–	97
Polynomial	10	3	0.5	–	–	92
Polynomial	10	4	2	–	–	96
Polynomial	10	4	4	–	–	95
Polynomial	10	4	1	–	–	93
Polynomial	10	4	0.5	–	–	92
Sigmoid	10	–	0.5	–	1	91
Sigmoid	10	–	0.2	–	1	92

Table 4.7 The training set results corresponding to both small and capital letters (for FSVM)

Kernel function	C	ρ	κ	σ	ϕ	Precision (%)
Linear	1	–	–	–	–	86
Linear	10	–	–	–	–	89
RBF	10	–	–	0.25	–	91
RBF	10	–	–	0.10	–	92
RBF	10	–	–	0.05	–	86
RBF	10	–	–	0.01	–	87
Polynomial	10	2	2	–	–	88
Polynomial	10	3	2	–	–	89
Polynomial	10	4	2	–	–	91
Sigmoid	10	–	0.5	–	–	92
Sigmoid	10	–	0.2	–	–	92

Table 4.8 The training set results corresponding to both small and capital letters (for FRSVM)

Kernel function	C	ρ	κ	σ	ϕ	Precision (%)
Linear	1	–	–	–	–	87
Linear	10	–	–	–	–	92
RBF	10	–	–	0.25	–	92
RBF	10	–	–	0.10	–	93
RBF	10	–	–	0.05	–	89
RBF	10	–	–	0.01	–	88
Polynomial	10	2	2	–	–	89
Polynomial	10	3	2	–	–	91
Polynomial	10	4	2	–	–	92
Sigmoid	10	–	0.5	–	–	93
Sigmoid	10	–	0.2	–	–	93

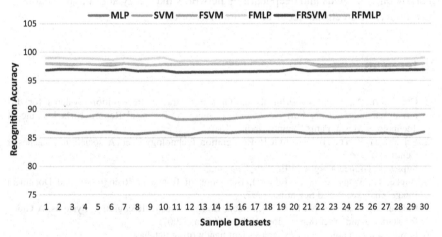

Fig. 4.10 The comparative performance of soft computing versus traditional techniques for English language

4.9 Further Discussions

All the feature based classification based methods used in this chapter give better results than the traditional approaches [1]. On comparing with other algorithms, it is observed that both FMLP and RFMLP work with high accuracy in almost all cases which include intersections of loop and instances of multiple crossings. This is mainly because these algorithms focussed on the processing of various asymmetries in the English characters. FMLP and RFMLP functions consider the conditions: (a) the validity of the algorithms for non-cursive French alphabets (b) the requirement of the algorithm that height of both upper and lower case characters to be proportionally same and (c) for extremely illegible handwriting the accuracy achieved by the algorithm is very less.

FSVM and FRSVM also give superior performance compared to traditional SVM [13]. FSVM and FRSVM achieve a precision rate up to 97% in case of training with sets corresponding to either small or capital letters and up to 98% in case of training with sets corresponding to both small and capital letters respectively. Thus the system achieved its goal through the recognition of characters from an image. The future research direction entails in expanding the system through addition of techniques which determine automatically the optimal parameters of kernel functions. Further, any version of SVM can be used to better perform the feature based classification task. We are also experimenting with other robust versions of SVM which will improve the overall recognition accuracy of the English characters.

Finally we conclude the chapter with a note that we are exploring further results on IAM dataset using other soft computing techniques such as fuzzy markov random fields and deep learning networks like fuzzy and rough versions of hierarchical bidirectional recurrent neural networks.

References

1. Chaudhuri, A., Some Experiments on Optical Character Recognition Systems for different Languages using Soft Computing Techniques, Technical Report, Birla Institute of Technology Mesra, Patna Campus, India, 2010.
2. Schantz, H. F., The History of OCR, Recognition Technology Users Association, Manchester Centre, VT, 1982.
3. https://en.wikipedia.org/wiki/English_language.
4. Bunke, H., Wang, P. S. P. (Editors), Handbook of Character Recognition and Document Image Analysis, World Scientific, 1997.
5. Cheriet, M., Kharma, N., Liu, C. L., Suen, C. Y., Character Recognition Systems: A Guide for Students and Practitioners, John Wiley and Sons, 2007.
6. http://www.iam.unibe.ch/fki/databases/iam-handwriting-database.
7. Chaudhuri, A., De, K., Job Scheduling using Rough Fuzzy Multi-Layer Perception Networks, Journal of Artificial Intelligence: Theory and Applications, 1(1), pp 4–19, 2010.
8. Chaudhuri, A., De, K., Chatterjee, D., Discovering Stock Price Prediction Rules of Bombay Stock Exchange using Rough Fuzzy Multi-Layer Perception Networks, Book Chapter: Forecasting Financial Markets in India, Rudra P. Pradhan, Indian Institute of Technology Kharagpur, (Editor), Allied Publishers, India, pp 69–96, 2009.
9. Pal, S. K., Mitra, S., Mitra, P., Rough-Fuzzy Multilayer Perception: Modular Evolution, Rule Generation and Evaluation, IEEE Transactions on Knowledge and Data Engineering, 15(1), pp 14–25, 2003.
10. Chaudhuri, A., Modified Fuzzy Support Vector Machine for Credit Approval Classification, AI Communications, 27(2), pp 189–211, 2014.
11. Chaudhuri, A., De, Fuzzy Support Vector Machine for Bankruptcy Prediction, Applied Soft Computing, 11(2), pp 2472–2486, 2011.

12. Chaudhuri, A., Fuzzy Rough Support Vector Machine for Data Classification, International Journal of Fuzzy System Applications, 5(2), pp 26–53, 2016.
13. Chaudhuri, A., Applications of Support Vector Machines in Engineering and Science, Technical Report, Birla Institute of Technology Mesra, Patna Campus, India, 2011.
14. Taghva, K., Borsack, J., Condit, A., Effects of OCR Errors on Ranking and Feedback using the Vector Space Model, Information Processing and Management, 32(3), pp 317–327, 1996.
15. Taghva, K., Borsack, J., Condit, A., Evaluation of Model Based Retrieval Effectiveness with OCR Text, ACM Transactions on Information Systems, 14(1), pp 64–93, 1996.
16. Taghva, K., Borsack, J., Condit, A., Erva, S., The Effects of Noisy Data on Text Retrieval, Journal of American Society for Information Science, 45 (1), pp 50–58, 1994.
17. Jain, A. K., Fundamentals of Digital Image Processing, Prentice Hall, India, 2006.
18. Russ, J. C., The Image Processing Handbook, CRC Press, 6th Edition, 2011.
19. Young, T. Y., Fu, K. S., Handbook of Pattern Recognition and Image Processing, Academic Press, 1986.
20. Gonzalez, R. C., Woods, R. E., Digital Image Processing, 3rd Edition, Pearson, 2013.
21. Jain, A. K., Duin, R. P. W., Mao, J., Statistical Pattern Recognition: A Review, IEEE Transactions on Pattern Analysis and Machine Intelligence, 22(1), pp 4–37, 2000.
22. De, R. K., Basak, J., Pal, S. K., Neuro-Fuzzy Feature Evaluation with Theoretical Analysis, Neural Networks, 12(10), pp 1429–1455, 1999.
23. De, R. K., Pal, N. R., Pal, S. K., Feature Analysis: Neural Network and Fuzzy Set Theoretic Approaches, Pattern Recognition, 30(10), pp 1579–1590, 1997.
24. https://www.csie.ntu.edu.tw/~cjlin/libsvm/.

Chapter 5
Optical Character Recognition Systems for French Language

Abstract The optical character recognition (OCR) systems for French language were the most primitive ones and occupy a significant place in pattern recognition. The French language OCR systems have been used successfully in a wide array of commercial applications. The different challenges involved in the OCR systems for French language is investigated in this chapter. The pre-processing activities such as text region extraction, skew detection and correction, binarization, noise removal, character segmentation and thinning are performed on the datasets considered. The feature extraction is performed through fuzzy Hough transform. The feature based classification is performed through important soft computing techniques viz rough fuzzy multilayer perceptron (RFMLP), fuzzy support vector machine (FSVM), fuzzy rough support vector machine (FRSVM) and hierarchical fuzzy bidirectional recurrent neural networks (HFBRNN). The superiority of soft computing techniques is demonstrated through the experimental results.

Keywords French language OCR · RFMLP · FSVM · FRSVM · HFBRNN

5.1 Introduction

In the present business scenario, several optical character recognition (OCR) systems have been developed for French language [7, 18]. French is the most widely spoken language across the globe [25] after English and Chinese. As a result of this, the development of OCR systems for French language occupy a significant place in pattern recognition [1]. French language OCR systems have been used successfully in a wide array of commercial applications [7]. The character recognition of French language has a high potential in data and word processing as the English language. Some commonly used applications of the OCR systems of French language [7] are automated postal address and ZIP code reading, data acquisition in bank checks, processing of archived institutional records etc.

© Springer International Publishing AG 2017
A. Chaudhuri et al., *Optical Character Recognition Systems for Different Languages with Soft Computing*, Studies in Fuzziness and Soft Computing 352, DOI 10.1007/978-3-319-50252-6_5

The standardization of OCR character set for French language was provided through ISO/IEC 8859 [25] as shown in Fig. 5.1. ISO/IEC 8859 is a joint ISO and IEC series of standards for 8-bit character encodings. The series of standards consists of numbered parts such as ISO/IEC 8859-1, ISO/IEC 8859-2 etc. There are 15 parts excluding the abandoned ISO/IEC 8859-12. ISO/IEC 8859 parts 1, 2, 3, and 4 were originally Ecma International standard ECMA-94. In the recent years, OCR for French language has gained a considerable momentum as the need for converting the scanned images into computer recognizable formats such as text documents has variety of applications. The French language based OCR systems is thus one of the most fascinating and challenging areas of pattern recognition with various practical applications [7].

The OCR process for any language involves extraction of defined characteristics called features to classify an unknown character into one of the known classes [1, 6, 7, 10] to a user defined accuracy level. As such any good OCR system is best defined in terms of the success of feature extraction and classification tasks. The same is true for French language. The process becomes tedious in case the language has dependencies where some characters look identical. Thus the classification becomes a big challenge.

Fig. 5.1 ISO/IEC 8859 font

In this chapter we start the investigation of OCR systems considering the different challenges involved in French language. The different pre-processing activities such as text region extraction, skew detection and correction, binarization, noise removal, character segmentation and thinning are performed on the considered datasets [26]. The feature extraction is performed through fuzzy Hough transform. The feature based classification is performed through important soft computing techniques viz rough fuzzy multilayer perceptron (RFMLP) [8, 9, 16] two support vector machine (SVM) based methods such as fuzzy support vector machine (FSVM) [4, 5] and fuzzy rough support vector machine (FRSVM) [3] and hierarchical fuzzy bidirectional recurrent neural networks (HFBRNN) [2]. The experimental results demonstrate the superiority of soft computing techniques over the traditional methods.

This chapter is structured as follows. In Sect. 5.2 a brief discussion about the French language script and datasets used for experiments are presented. The different challenges of OCR for French language are highlighted in Sect. 5.3. The next section illustrates the data acquisition. In Sect. 5.5 different pre-processing activities on the datasets such as text region extraction, skew detection and correction, binarization, noise removal, character segmentation and thinning are presented. This is followed by a discussion of feature extraction on French language dataset in Sect. 5.6. The Sect. 5.7 explains the state of art of OCR for French language in terms of feature based classification methods. The corresponding experimental results are given in Sect. 5.8. Finally Sect. 5.9 concludes the chapter with some discussions and future research directions.

5.2 French Language Script and Experimental Dataset

In this section we present brief information about the French language script and the dataset used for experiments. French is a romance language of Indo-European family [25]. The language descended from the Vulgar Latin of the Roman Empire. French has evolved from Gallo-Romance, the spoken Latin in Gaul, more specifically in Northern Gaul. Its closest relatives are the other languages spoken in northern France and in southern Belgium which French has largely supplanted. French was also influenced by native Celtic languages of Northern Roman Gaul like Gallia Belgica and by the Germanic Frankish language of the post-Roman Frankish invaders. Today, owing to France's past overseas expansion there are numerous French based creole languages most notably Haitian Creole. French is the official language in 29 countries, most of which are members of *la francophonie*, the community of French-speaking countries. It is spoken as a first language in France, southern Belgium, western Switzerland, Monaco, certain parts of Canada and the United States and by various communities elsewhere. As of 2016, 40% of the francophone population is in Europe, 35% in sub-Saharan Africa, 15% in North Africa and the Middle East, 8% in the Americas and 1% in Asia and Oceania.

French is the fourth most widely spoken mother tongue in the European Union. About 1/5 of non-Francophone Europeans speak French. As a result of French and Belgian colonialism from the 17th and 18th century onward French was introduced to new territories in the Americas, Africa and Asia. In 2016, French was estimated to have 77–110 million native speakers and 190 million secondary speakers. Approximately 270 million people speak the French language. French has a long history as an international language of commerce, diplomacy, literature, and scientific standards and is an official language of many international organisations including United Nations, European Union, NATO, WTO, the International Olympic Committee and the ICRC.

The French language dataset used for performing OCR experiments is the IRESTE IRONFF dataset and is adapted from [26]. IRESTE IRONFF database contains French text which is used here to for training and testing. The database contains unconstrained handwritten text which are scanned at a resolution of 300 dpi and saved as PNG images with 256 gray levels. The Fig. 5.2 shows a sample snapshot from the database. IRESTE IRONFF database is structured as follows:

(a) 696 writers contributed samples of their handwriting
(b) 4086 isolated digits

nonbreaking space	¡	¢	£	¤	¥	¦	§	¨	©	ª	«
	¡ ¡	¢ ¢	£ £	¤ ¤	¥ ¥	¦ ¦	§ §	¨ ¨	© ©	ª ª	« «
¬	-	®		°	±	²	³	´	µ	¶	·
¬ ¬	­ ­	® ®	¯ ¯	° °	± ±	² ²	³ ³	´ ´	µ µ	¶ ¶	· ·
¸	¹	º	»	¼	½	¾	¿	À	Á	Â	Ã
¸ ¸	¹ ¹	º º	» »	¼ ¼	½ ½	¾ ¾	¿ ¿	À À	Á Á	Â Â	Ã Ã
Ä	Å	Æ	Ç	È	É	Ê	Ë	Ì	Í	Î	Ï
Ä Ä	Å Å	Æ &Aelig;	Ç Ç	È È	É É	Ê Ê	Ë Ë	Ì Ì	Í Í	Î Î	Ï Ï
Ð	Ñ	Ò	Ó	Ô	Õ	Ö	×	Ø	Ù	Ú	Û
Ð Ð	Ñ Ñ	Ò Ò	Ó Ó	Ô Ô	Õ Õ	Ö Ö	× ×	Ø Ø	Ù Ù	Ú Ú	Û Û
Ü	Ý	Þ	ß	à	á	â	ã	ä	å	æ	ç
Ü Ü	Ý Ý	Þ Þ	ß ß	à à	á á	â â	ã ã	ä ä	å å	æ æ	ç ç
è	é	ê	ë	ì	í	î	ï	ð	ñ	ò	ó
è è	é é	ê ê	ë ë	ì ì	í í	î î	ï ï	ð ð	ñ ñ	ò ò	ó ó
ô	õ	ö	÷	ø	ù	ú	û	ü	ý	þ	ÿ
ô ô	õ õ	ö ö	÷ ÷	ø ø	ù ù	ú ú	û û	ü ü	ý ý	þ þ	ÿ ÿ

Fig. 5.2 A sample text snapshot from IRESTE IRONFF database

(c) 10,685 isolated lower case letters
(d) 10,679 isolated upper case letters
(e) 410 EURO signs
(f) 31,346 isolated words from a 197 word lexicon with 28,657 French words.

Further details are available at [26].

5.3   Challenges of Optical Character Recognition Systems for French Language

After English language the OCR for French language has become one of the most successful applications of technology in pattern recognition and artificial intelligence. The OCR for French language has been the topic of active research since past few decades [7]. The most commercially available OCR system for French language is ABBYY FineReader [27]. Considering the important aspects of versatility, robustness and efficiency, the commercial OCR systems are generally divided into four generations [7] as highlighted in Chap. 2. It is to be noted that this categorization refers also to the OCRs of French language.

Despite decades of research and existence of established commercial OCR products based on French language, the output from such OCR processes often contains errors. The more highly degraded is input, the greater is error rate. Since inputs form the first stage in a pipeline where later stages are designed to support sophisticated information extraction and exploitation applications, it is important to understand the effects of recognition errors on downstream analysis routines. Few questions are required to be addressed in this direction. They are as follows:

(a) Are all recognition errors equal in impact or some are worse than others?
(b) Can the performance of each stage be optimized in isolation or the end-to-end system should be considered?
(c) In balancing the trade-off between the risk of over and under segmenting characters during OCR where should the line be drawn to maximize overall performance?

The answers to these questions often influence the way OCR systems for French language are designed and build for analysis [7].

The French language OCR system converts numerous published books in French language into editable computer text files. The latest research in this area has grown to incorporate some new methodologies to overcome the complexity of French writing style. All these algorithms have still not been tested for complete characters of French alphabet. Hence, there is a quest for developing an OCR system which handles all classes of French text and identify characters among these classes increasing versatility, robustness and efficiency in commercial OCR systems. The recognition of printed French characters is itself a challenging problem since there is a variation of the same character due to change of fonts or introduction of different types of noises. There may be noise pixels that are introduced due to scanning of the image. A significant amount of research has been done towards text data processing in French language from noisy sources [7]. The majority of the work has focused predominately on errors that arise during speech recognition systems [21, 22]. Several research papers have appeared which examines the noise problem from variety of perspectives with most emphasizing issues that are inherent in written and spoken French language [7]. However, there has been less work concentrating on noise induced by OCR. Some earlier works by [19] show that moderate error rates have little impact on effectiveness of traditional information retrieval measures. However, this conclusion is tied to certain assumptions about information retrieval through bag of words, OCR error rate which may not be too high and length of documents which may not be too short. Some other notable research works in this direction are given in [20, 21]. All these works try to address some significant issues involved in OCR systems for French language such as error correction, performance evaluation etc. involving flexible and rigorous mathematical treatment [7]. Besides this any French character can be represented in variety of fonts and sizes as per the needs and requirements of application. Further the character with same font and size may also be bold face character as well as normal one [9]. Thus the width of stroke also significantly affects recognition process. Therefore, a good character recognition approach for French language [7, 15, 17, 22]:

(a) Must eliminate noise after reading binary image data
(b) Smooth image for better recognition
(c) Extract features efficiently
(d) Train the system and
(e) Classify patterns accordingly.

5.4 Data Acquisition

The progress in automatic character recognition systems in French language is motivated from two categories according to the mode of data acquisition which can be either online or offline character recognition systems [7]. Following the lines of English language, the data acquisition of French language can be either online or offline character recognition systems. The offline character recognition captures data from paper through optical scanners or cameras whereas online recognition systems utilize digitizers which directly capture writing with the order of strokes, speed, pen-up and pen-down information. As such the scope of this text is restricted to OCR systems, we confine our discussion to offline character recognition [7] for French language. The French language datasets used in this research is mentioned in Sect. 5.2.

5.5   Data Pre-processing

Once the data has been acquired properly we proceed to pre-process the data. In pre-processing stage [13, 14, 17] a series of operations are performed here which includes text region extraction, skew detection and correction, binarization, noise removal, character segmentation and thinning or skeletonization. The main objective of pre-processing is to organize information so that the subsequent character recognition task becomes simpler. It essentially enhances the image rendering it suitable for segmentation.

5.5.1   Text Region Extraction

The text region extraction is used here as the first step in character recognition process. The input image $I_{P \times Q}$ is first partitioned into m number of blocks $B_i; i = 1, 2, \ldots \ldots, m$ such that:

$$B_i \cap B_j = \emptyset \tag{5.1}$$

$$I_{P \times Q} = \bigcup_{i=1}^{m} B_i \tag{5.2}$$

A block B_i is a set of pixels represented as $B_i = [f(x, y)]_{H \times W}$ where H and W are the height and the width of the block respectively. Each individual block B_i is classified as either information block or background block based on the intensity variation within it. After removal of background blocks adjacent or contiguous information blocks constitute isolated components called as regions $R_i; i = 1, 2, \ldots, n$ such that:

$$R_i \cap R_j = \emptyset \, \forall \text{ but } \bigcup_{i=1}^{n} R_i \neq I_{P \times Q} \tag{5.3}$$

This is because some background blocks have been removed. The area of a region is always a multiple of the area of the blocks. These regions are then classified as text region or non-text region using various characteristics features of textual and non-textual regions such as dimensions, aspect ratio, information pixel density, region area, coverage ratio, histogram, etc. A detail description of this technique has been presented in [7]. The Fig. 5.3 shows a camera captured image and the text regions extracted from it.

Fig. 5.3 The camera captured image and the text regions extracted from it

5.5.2 Skew Detection and Correction

When a text document is fed into scanner either mechanically or manually a few degrees of tilt or skew is unavoidable. In skew angle the text lines in digital image make angle with horizontal direction. A number of methods are available in literature for identifying image skew angles [1]. They are basically categorized on the basis of projection profile analysis, nearest neighbor clustering, Hough transform, cross correlation and morphological transforms. The camera captured images very often suffer from skew and perspective distortion [15]. They occur due to non-parallel axes or planes at the time of capturing the image. The acquired image does not become uniformly skewed mainly due to perspective distortion. The skewness of different portions of the image may vary between $+\alpha$ to $-\beta$ degrees where both α and β are positive numbers. Hence, the image cannot be deskewed at a single pass. On the other hand the effect of perspective distortion is distributed throughout the image. Its effect is hardly visible within a small region for example, the area of a character of the image. At the same time the image segmentation module generates only a few text regions. These text regions are deskewed using a computationally efficient and fast skew correction technique presented in [7].

Every text region basically has two types of pixels viz dark and gray. The dark pixels constitute the texts and the gray pixels are background around the texts. For the four sides of virtual bounding rectangle of a text region, there are four sets of values known as profiles. If the length and breadth of the bounding rectangle are M and N respectively, then two profiles will have M values each and the other two will have N values each. These values are the distances in terms of pixel from a side to the first gray or black pixel of the text region. Among these four profiles, the one which is from the bottom side of the text region is taken into consideration for estimating skew angle as shown in Fig. 5.4. This bottom profile is denoted as $\{h_i; i = 1, 2, \ldots\ldots, M\}$.

The mean $\mu = \frac{1}{M} \sum_{i=1}^{M} h_i$ and the first order moment $\tau = \frac{1}{M} \sum_{i=1}^{M} \mu - h_i$ values are calculated. Then, the profile size is reduced by excluding some h_i

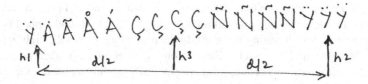

Fig. 5.4 The calculation of skew angle from *bottom* profile of a text region

values that are not within the range $\mu \pm \tau$. The central idea behind this exclusion is that these elements hardly contribute to the actual skew of the text region. Now from the remaining profile, the elements viz leftmost h_1, rightmost h_2 and the middle one h_3 are chosen. The final skew angle is computed by averaging the three skew angles obtained from three pairs $h_1 - h_3$, $h_3 - h_2$ and $h_1 - h_2$. Once the skew angle for a text region is estimated it is rotated by the same angle.

5.5.3 Binarization

Binarization [13, 17] is the next step in character recognition process. A large number of binarization techniques are available in the literature [7] each of which is appropriate to particular image types. Its goal is to reduce the amount of information present in the image and keep only the relevant information. Generally the binarization techniques of gray scale images are classified into two categories viz overall threshold where single threshold is used in the entire image to form two classes (text and background) and local threshold where values of thresholds are determined locally (pixel-by-pixel or region-by-region). Here the skew corrected text region is binarized using an efficient binarization technique [22]. The algorithm is given below:

> **Binarization Algorithm:**
> **begin**
>   **for** all pixels in (x, y) in *Text Region*
>   **if** intensity $(x, y) < (G_{max} + G_{min})/2$
>   **then** mark (x, y) as foreground
>   **else if** number of foreground neighbors > 4
>   **then** mark (x, y) as foreground
>   **else** mark (x, y) as background
>   **end if**
>   **end if**
>   **end for**
> **end**

This is an improved version of bernsen's binarization method [7]. The arithmetic mean of maximum G_{max} and minimum G_{min} gray levels around a pixel is taken as the threshold for binarizing the pixel. In the present algorithm the eight immediate neighbors around the pixel subject to binarization are also taken as deciding factors for binarization. This approach is especially useful to connect the disconnected foreground pixels of a character.

5.5.4 Noise Removal

The scanned text documents often contain noise that arises due to printer, scanner, print quality, document age etc. Therefore, it is necessary to filter noise [13] before the image is processed. Here a low-pass filter is used to process the image [15] which is used for later processing. The main objective in the design of a noise filter is that it should remove as much noise as possible while retaining the entire signal [17].

5.5.5 Character Segmentation

Once the text image is skew corrected, binarized and noise removed, the actual text content is extracted. This process leads to character segmentation [13]. The commonly used segmentation algorithms in this direction are connected component labeling, x-y tree decomposition, run length smearing and Hough transform [7]. After binarizing a noise free text region, the horizontal histogram profile $\{f_i; i = 1, 2, \ldots\ldots, H_R\}$ of the region as shown in Fig. 5.5a, b is analyzed

Fig. 5.5 **a** The horizontal histogram of text regions for their segmentation (skewed text region and its horizontal histogram). **b** The horizontal histogram of text regions for their segmentation (skew corrected text region and its horizontal histogram)

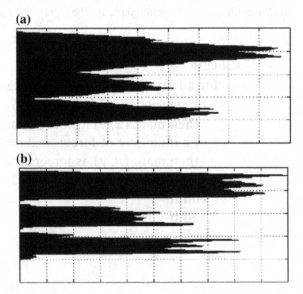

for segmenting the region into text lines. Here f_i denotes the number of black pixel along ith of the text region and hidden region denotes the height of the deskewed text region. Firstly, all possible line segments are determined by thresholding the profile values. The threshold is chosen so as to allow over segmentation. Text line boundaries are referred by the values of i for which the value of f_i is less than the threshold. Thus, n such segments represent $n - 1$ text lines. After that the inter segment distances are analyzed and some segments are rejected based on the idea that the distance between two lines in terms of pixels will not be too small and the inter-segment distances are likely to become equal. A detail description of the method is given in [15]. Using vertical histogram profile of each individual text lines, words and characters are segmented. Sample segmented characters have been shown in Fig. 5.6a, b.

Fig. 5.6 a Skew correction and segmentation of text regions (an extracted text region). **b** Skew correction and segmentation of text regions (characters segmented from de-skewed text region)

Fig. 5.7 An image before and after thinning

5.5.6 Thinning

The character segmentation process is followed by thinning or skeletonization. In thinning one-pixel-width representation or skeleton of an object is obtained by preserving connectedness of the object and its end points [17]. The thinning process reduces image components to their essential information so that further analysis and recognition are facilitated. For instance, an alphabet can be handwritten with different pens giving different stroke thicknesses but information presented is same. This enables easier subsequent detection of pertinent features. As an illustration consider letter ë shown in Fig. 5.7 before and after thinning. A number of thinning algorithms have been used in the past with considerable success. The most common algorithm used is the classical hilditch algorithm and its variants [15]. Here hilditch algorithm is used for thinning [7]. For recognizing large graphical objects with filled regions which are often found in logos, region boundary detection is useful but for small regions corresponding to individual characters neither thinning nor boundary detection is performed. Rather entire pixel array representing the region is forwarded to subsequent stage of analysis.

5.6 Feature Extraction Through Fuzzy Hough Transform

As mentioned in Chap. 4 the heart of any OCR system is the formation of feature vector used in recognition stage. This fact is also valid for French language OCR system. This phase extracts the features from segmented areas of image containing characters to be recognized that distinguishes an area corresponding to a letter from an area corresponding to other letters. The feature extraction phase can thus be visualised as finding a set of parameters or features that define the shape of character as precise and unique. In Chap. 3 the term feature extraction is often used synonymously by feature selection which refers to algorithms that select the best subset of input feature set. These methods create new features based on transformations or combination of original features [7]. The features selected help in discriminating the characters. Achieving high recognition performance is attributed towards the selection of appropriate feature extraction methods. A large number of OCR based feature extraction methods are available in literature [7] except that the selected method depends on the application concerned. There is no universally accepted set of feature vectors in OCR. The features that capture topological and geometrical shape information are the most desired ones. The features that capture spatial distribution of black (text) pixels are also very important [13]. The Hough transform based feature extraction approach is successfully applied for the OCR of French language [7].

The Hough transform method is used for detection of lines and curves from images [15]. The basic Hough transform is generalized through fuzzy probabilistic concepts [7]. The fuzzy Hough transform treats image points as fuzzy points. The Hough transform for line detection uses mapping $r = x cos\theta + y sin\theta$ which provides three important characteristics of line in an image pattern. The parameters r and θ specify position and orientation of line. The count of (r, θ) accumulator cell used in Hough transform implementation specifies number of black pixels lying on it. With this motivation, a number of fuzzy sets on (r, θ) accumulator cells are defined. Some important fuzzy set definitions used here are presented in Table 5.1 for θ values in first quadrant. The definitions are extended for other values of θ. The fuzzy sets viz. long_line and short_line extract length information of different

Table 5.1 Fuzzy set membership functions defined on Hough transform accumulator cells for line detection (x and y denote height and width of each character pattern)

Fuzzy set	Membership function
long_line	$\frac{cellcount}{\sqrt{x^2+y^2}}$
short_line	2(long_line) if $count \leq \sqrt{x^2 + y^2}/2$ 2(1—long_line) if $count > \sqrt{x^2 + y^2}/2$
nearly_horizontal_line	$\frac{\theta}{90}$
nearly_vertical_line	1—nearly_horizontal_line
slant_line	2(nearly_horizontal_line) if $\theta \leq 45$ 2(1—nearly_horizontal_line) if $\theta > 45$
near_top	r/x if nearly_horizontal_line > nearly_vertical_line 0 otherwise
near_bottom	(1—near_top) if nearly_horizontal_line > nearly_vertical_line 0 otherwise
near_vertical_centre	2(near_top) if ((nearly_horizontal_line > nearly_vertical_line) and ($r \leq x/2$)) 2(1—near_top) if ((nearly_horizontal_line > nearly_vertical_line) and ($r > x/2$)) 0 otherwise
near_right_border	r/y if nearly_vertical_line > nearly_horizontal_line 0 otherwise
near_left_border	(1—near_right_border) if nearly_vertical_line > nearly_horizontal_line 0 otherwise
near_horizontal_centre	2(near_right_border) if ((nearly_vertical_line > nearly_horizontal_line) and ($r \leq y/2$)) 2(1—near_right_border) if ((nearly_vertical_line > nearly_horizontal_line) and ($r > y/2$)) 0 otherwise

lines in pattern. The nearly_horizontal, nearly_vertical and slant_line represent skew and near_top, near_bottom, near_vertical_centre, near_right, near_left and near_horizontal_centre extract position information of these lines. The characteristics of different lines in an image pattern are mapped into properties of these fuzzy sets. For interested readers further details are available in [7].

Based on basic fuzzy sets [23, 24], fuzzy sets are further synthesized to represent each line in pattern as combination of its length, position and orientation using t-norms [24]. The synthesized fuzzy sets are defined as long_slant_line $\equiv$ t-norm (slant_line, long_line), short_slant_line $\equiv$ t-norm (slant_line, short_line), nearly_vertical_long_line_near_left $\equiv$ t-norm (nearly_vertical_line, long_line, near_left_border). Similar basic fuzzy sets such as large_circle, dense_circle, centre_near_ top etc. and synthesized fuzzy sets such as small_dense_circle_near_top, large_dense_circle_near_centre etc. are defined on (p, q, t) accumulator cells for circle extraction using Hough transform $t = \sqrt{(x - p)^2 + (y - q)^2}$. For a circle extraction (p, q) denotes origin, c is radius and count specifies the number of pixels lying on circle. A number of t-norms are available as fuzzy intersections among which standard intersection t-norm $(p, q) \equiv \min (p, q)$. For other pattern recognition problems suitable fuzzy sets may be similarly synthesized from basic sets of fuzzy Hough transform. A non-null support of synthesized fuzzy set implies presence of corresponding feature in a pattern. The height of each synthesized fuzzy set is chosen to define feature element and set of n such feature elements constitute an n-dimensional feature vector for a character.

5.7 Feature Based Classification: Sate of Art

After extracting the essential features from the pre-processed character image the concentration pointer turns on the feature based classification methods for OCR of French language. Considering the wide array of feature based classification methods for OCR, these methods are generally grouped into following four broad categories [11, 12, 20]:

(a) Statistical methods
(b) ANN based methods
(c) Kernel based methods
(d) Multiple classifier combination methods.

We discuss here the work done on French characters using some of the above-mentioned classification methods. All the classification methods used here are soft computing based techniques. The next three subsections highlight the feature based classification based on RFMLP [8, 9, 16], FSVM [4, 5], FRSVM [3] and HFBRNN [2] techniques. These methods are already discussed in Chap. 3. For further details interested readers can refer [6, 7].

5.7.1 Feature Based Classification Through Rough Fuzzy Multilayer Perceptron

The French language characters are recognized here through RFMLP [8, 9, 16] which is discussed in Chap. 3. For the French language RFMLP has been used with considerable success [7]. The diagonal feature extraction scheme is used for drawing of features from the characters. For classification stage we use these features extracted. RFMLP used here is a feed forward network with back propagation ANN having four hidden layers. The architecture used is 70–100–100–100–100–100–52 [7] for classification. The four hidden layers and output layer uses tangent sigmoid as the activation function. The feature vector is again denoted by $F = (f_1, \ldots \ldots, f_m)$ where m denotes number of training images and each f has a length of 70 which represent the number of input nodes. The 52 neurons in the output layer correspond to the 26 French alphabets both uppercase and lowercase [26]. The network training parameters are briefly summarized as [7]:

(a) Input nodes: 70
(b) Hidden nodes: 100 at each layer
(c) Output nodes: 52
(d) Training algorithm: Scaled conjugate gradient backpropagation
(e) Performance function: Mean Square Error
(f) Training goal achieved: 0.000002
(g) Training epochs: 4000.

5.7.2 Feature Based Classification Through Fuzzy and Fuzzy Rough Support Vector Machines

In similar lines to the English language character recognition through FSVM and FRSVM in Sect. 4.7.3, FSVM [4, 5] and FRSVM [3] discussed in Chap. 3 is used here to recognize the French language characters. Over the past years SVM based methods have shown a considerable success in feature based classification [6]. With this motivation FSVM and FRSVM are here for feature classification task. Both FSVM and FRSVM offer the possibility of selecting different types of kernel functions such as sigmoid, RBF, linear functions and determining the best possible values for these kernel parameters [4, 5, 6]. After selecting the kernel type and its parameters, FSVM and FRSVM are trained with the set of features obtained from other phases. Once the training gets over, FSVM and FRSVM are used to classify new character sets. The implementation is achieved through LibSVM library [28].

5.7.3 Feature Based Classification Through Hierarchical Fuzzy Bidirectional Recurrent Neural Networks

After RFMLP, FSVM and FRSVM, the French language characters are recognized through HFBRNN [2] which is discussed in Chap. 3. For the French language HFBRNN has been used with considerable success [7]. HBRNN is a deep learning based technique [2]. In the recent past deep learning based techniques have shown a considerable success in feature based classification [2]. With this motivation HBRNN is used here for feature classification task and takes full advantage of deep recurrent neural network (RNN) towards modeling long-term information of data sequences. The recognition of characters done by HFBRNN at review level. The performance of HFBRNN is improved by fine tuning parameters of the network in a hierarchical fashion. The motivation is obtained from long short term memory (LSTM) [2] and bidirectional LSTM (BLSTM) [2]. The evaluation is done on different types of highly biased character data. The implementation of HBRNN is performed in MATLAB [7].

5.8 Experimental Results

In this section, the experimental results for soft computing tools viz RFMLP, FSVM, FRSVM and HFBRNN on the French language dataset [26] highlighted in Sect. 5.2 are presented. The prima face is to select the best OCR system for French language [7].

5.8.1 Rough Fuzzy Multilayer Perceptron

The experimental results for a subset of the French characters both uppercase and lowercase using RFMLP are shown in the Table 5.2. Like the English characters, the French characters can be categorized as ascenders (characters such as ç, Ç etc.) and descenders (characters such as þ, Á etc.). RFMLP shows better results than the traditional methods [7]. The testing algorithm is applied to all standard cases of characters [26] in samples. An accuracy of around 99% has been achieved in all the cases. However, after successful training and testing of algorithm the following flaws are encountered [7] which are identical to those encountered in English language:

(a) There may an instance when there is an occurrence of large and disproportionate symmetry in both ascenders as well as descenders as shown in Fig. 5.8a, b.

(b) There may be certain slants in certain characters which results in incorrect detection of characters as shown in Fig. 5.9.

Table 5.2 The experimental results for a subset of the French characters using RFMLP

Characters	Successful recognition (%)	Unsuccessful recognition (%)	No recognition (%)
á	99	1	0
â	99	1	0
ã	99	1	0
ä	99	1	0
å	99	1	0
ç	99	1	0
è	98	1	1
é	99	1	0
ê	99	1	0
ë	99	1	0
ì	99	1	0
í	99	1	0
î	98	1	1
ï	98	1	1
ò	99	1	0
þ	99	1	0
û	99	1	0
ú	99	1	0
ù	99	1	0
ó	99	1	0
ô	99	1	0
õ	99	1	0
ö	98	1	1
ñ	98	1	1
ý	99	1	0
ÿ	99	1	0
Á	99	1	0
Â	99	1	0
Ã	99	1	0
Ä	99	1	0
Å	99	1	0
Ç	99	1	0
È	99	1	0
É	99	1	0
Ê	99	1	0
Ë	99	1	0
Ì	99	1	0
Í	99	1	0
Î	98	1	1
Ï	98	1	1

(continued)

Table 5.2 (continued)

Characters	Successful recognition (%)	Unsuccessful recognition (%)	No recognition (%)
Ò	99	1	0
Ó	99	1	0
Ô	99	1	0
Õ	99	1	0
Ö	99	1	0
Ù	99	1	0
Ú	99	1	0
Ü	99	1	0
Ý	98	1	1
þ	98	1	1
Đ	99	1	0
Ñ	99	1	0

Fig. 5.8 The disproportionate symmetry in characters 'þ' and 'ï'

(a)   **(b)**

Fig. 5.9 The character 'ï' with slant

Fig. 5.10 The uneven pixel combination for character 'È' (region-i on the *left image* has less white space than the *right image*)

(c) Sometimes one side of the image may have less white space and more pixel concentration whereas other side may have more white space and less pixel concentration. Due to this some characters are wrongly detected as shown in Fig. 5.10.

5.8.2 Fuzzy and Fuzzy Rough Support Vector Machines

FSVM and FRSVM show promising results for feature classification task through different types of kernel functions by selecting the best possible values for kernel parameters [3, 4, 5]. For testing accuracy of the system in the first test case scenario we use an image which contained 100 small letters as shown in Fig. 5.11. The training set is constructed through two images containing 40 examples of each small letter in the French alphabet which took around 19.69 s. The results are presented in Tables 5.3 and 5.4. The parameter C is regularization parameter, ρ is bias term, κ is kernel function, σ is smoothing function which reduces variance and ϕ is mapping function in feature space. Further discussion on these parameters is available in [6].

For the next test case scenario we use for training only the features corresponding to capital letters. The image used for testing contained 100 letters as shown in

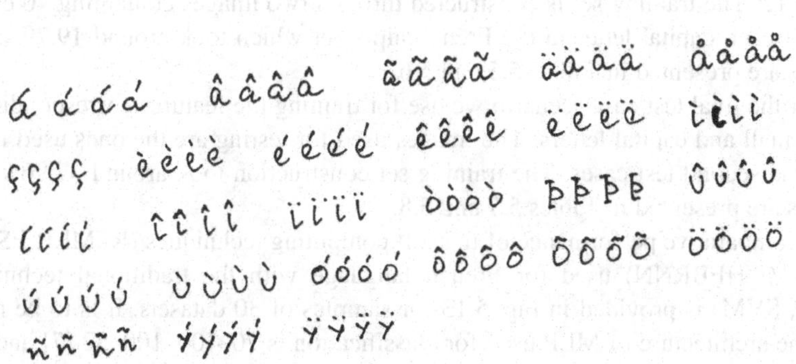

Fig. 5.11 The test image for small letters

Table 5.3 The training set results corresponding to small letters (for FSVM)

Kernel function	C	ρ	κ	σ	ϕ	Precision (%)
Linear	1	–	–	–	–	93
Linear	10	–	–	–	–	93
Linear	100	–	–	–	–	93
RBF	10	–	–	0.25	–	94
RBF	10	–	–	0.15	–	94
RBF	10	–	–	0.10	–	95
RBF	10	–	–	0.05	–	95
RBF	10	–	–	0.03	–	96
RBF	10	–	–	0.02	–	96
RBF	10	–	–	0.01	–	97
RBF	10	–	–	0.005	–	97
Polynomial	10	2	2	–	–	96
Polynomial	10	2	4	–	–	96
Polynomial	10	2	1	–	–	96
Polynomial	10	2	0.5	–	–	94
Polynomial	10	3	2	–	–	93
Polynomial	10	3	4	–	–	93
Polynomial	10	3	1	–	–	95
Polynomial	10	3	0.5	–	–	95
Polynomial	10	4	2	–	–	95
Polynomial	10	4	4	–	–	93
Polynomial	10	4	1	–	–	95
Polynomial	10	4	0.5	–	–	95
Sigmoid	10	–	0.5	–	1	93
Sigmoid	10	–	0.5	–	5	93
Sigmoid	10	–	0.2	–	1	94
Sigmoid	10	–	0.7	–	1	95

Fig. 5.12. The training set is constructed through two images containing 40 examples of each capital letter in the French alphabet which took around 19.79 s. The results are presented in Tables 5.5 and 5.6.

For the final test case scenario we use for training the features corresponding to both small and capital letters. The images used for testing are the ones used in the first and second test cases. The training set construction took around 37.96 s. The results are presented in Tables 5.7 and 5.8.

A comparative performance of the soft computing techniques (RFMLP, FSVM, FRSVM, HFBRNN) used for French language with the traditional techniques (MLP, SVM) is provided in Fig. 5.13 for samples of 30 datasets. It is to be noted that the architecture of MLP used for classification is 70–100–100–52 [7] and sigmoid kernel is used with SVM.

Table 5.4 The training set results corresponding to small letters (for FRSVM)

Kernel function	C	ρ	κ	σ	ϕ	Precision (%)
Linear	1	–	–	–	–	94
Linear	10	–	–	–	–	93
Linear	100	–	–	–	–	93
RBF	10	–	–	0.25	–	95
RBF	10	–	–	0.15	–	95
RBF	10	–	–	0.10	–	96
RBF	10	–	–	0.05	–	96
RBF	10	–	–	0.03	–	97
RBF	10	–	–	0.02	–	97
RBF	10	–	–	0.01	–	98
RBF	10	–	–	0.005	–	98
Polynomial	10	2	2	–	–	98
Polynomial	10	2	4	–	–	98
Polynomial	10	2	1	–	–	98
Polynomial	10	2	0.5	–	–	95
Polynomial	10	3	2	–	–	93
Polynomial	10	3	4	–	–	93
Polynomial	10	3	1	–	–	95
Polynomial	10	3	0.5	–	–	95
Polynomial	10	4	2	–	–	95
Polynomial	10	4	4	–	–	94
Polynomial	10	4	1	–	–	95
Polynomial	10	4	0.5	–	–	95
Sigmoid	10	–	0.5	–	1	95
Sigmoid	10	–	0.5	–	5	95
Sigmoid	10	–	0.2	–	1	96
Sigmoid	10	–	0.7	–	1	96

All the tests are conducted on PC having Intel P4 processor with 4.43 GHz, 512 MB DDR RAM @ 400 MHz with 512 KB cache. The training set construction was the longest operation of the system where the processor was loaded to 25% and the application occupied around 54.16 MB of memory. During idle mode the application consumes 43.02 MB of memory.

5.8.3 Hierarchical Fuzzy Bidirectional Recurrent Neural Networks

The experimental results for a subset of the French characters both uppercase and lowercase using HFBRNN are shown in the Table 5.9. The French characters can

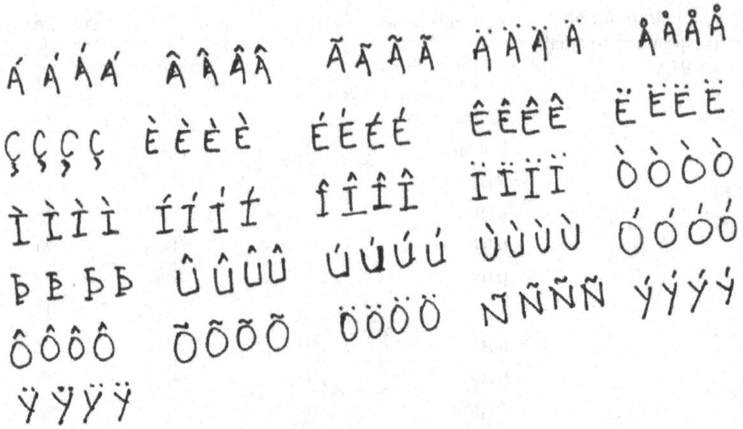

Fig. 5.12   The test image for capital letters

Table 5.5   The training set results corresponding to capital letters (for FSVM)

Kernel function	C	ρ	κ	σ	ϕ	Precision (%)
Linear	1	–	–	–	–	94
Linear	10	–	–	–	–	95
Linear	100	–	–	–	–	93
RBF	10	–	–	0.25	–	96
RBF	10	–	–	0.15	–	95
RBF	10	–	–	0.10	–	97
RBF	10	–	–	0.05	–	94
RBF	10	–	–	0.03	–	93
RBF	10	–	–	0.02	–	93
RBF	10	–	–	0.01	–	94
RBF	10	–	–	0.005	–	94
Polynomial	10	2	2	–	–	89
Polynomial	10	2	4	–	–	94
Polynomial	10	2	1	–	–	93
Polynomial	10	2	0.5	–	–	94
Polynomial	10	3	2	–	–	94
Polynomial	10	3	4	–	–	89
Polynomial	10	3	1	–	–	96
Polynomial	10	3	0.5	–	–	93
Polynomial	10	4	2	–	–	95
Polynomial	10	4	4	–	–	94
Polynomial	10	4	1	–	–	93
Polynomial	10	4	0.5	–	–	93
Sigmoid	10	–	0.5	–	1	89
Sigmoid	10	–	0.2	–	1	89

Table 5.6 The training set results corresponding to capital letters (for FRSVM)

Kernel Function	C	ρ	κ	σ	ϕ	Precision (%)
Linear	1	–	–	–	–	95
Linear	10	–	–	–	–	95
Linear	100	–	–	–	–	93
RBF	10	–	–	0.25	–	97
RBF	10	–	–	0.15	–	96
RBF	10	–	–	0.10	–	98
RBF	10	–	–	0.05	–	95
RBF	10	–	–	0.03	–	94
RBF	10	–	–	0.02	–	93
RBF	10	–	–	0.01	–	93
RBF	10	–	–	0.005	–	95
Polynomial	10	2	2	–	–	93
Polynomial	10	2	4	–	–	95
Polynomial	10	2	1	–	–	94
Polynomial	10	2	0.5	–	–	95
Polynomial	10	3	2	–	–	96
Polynomial	10	3	4	–	–	94
Polynomial	10	3	1	–	–	97
Polynomial	10	3	0.5	–	–	93
Polynomial	10	4	2	–	–	96
Polynomial	10	4	4	–	–	95
Polynomial	10	4	1	–	–	93
Polynomial	10	4	0.5	–	–	93
Sigmoid	10	–	0.5	–	1	94
Sigmoid	10	–	0.2	–	1	93

Table 5.7 The training set results corresponding to both small and capital letters (for FSVM)

Kernel Function	C	ρ	κ	σ	ϕ	Precision (%)
Linear	1	–	–	–	–	86
Linear	10	–	–	–	–	89
RBF	10	–	–	0.25	–	93
RBF	10	–	–	0.10	–	93
RBF	10	–	–	0.05	–	86
RBF	10	–	–	0.01	–	89
Polynomial	10	2	2	–	–	88
Polynomial	10	3	2	–	–	89
Polynomial	10	4	2	–	–	93
Sigmoid	10	–	0.5	–	–	93
Sigmoid	10	–	0.2	–	–	93

Table 5.8 The training set
results corresponding to both
small and capital letters (for
FRSVM)

Kernel function	C	ρ	κ	σ	ϕ	Precision (%)
Linear	1	–	–	–	–	87
Linear	10	–	–	–	–	93
RBF	10	–	–	0.25	–	93
RBF	10	–	–	0.10	–	94
RBF	10	–	–	0.05	–	89
RBF	10	–	–	0.01	–	89
Polynomial	10	2	2	–	–	89
Polynomial	10	3	2	–	–	94
Polynomial	10	4	2	–	–	93
Sigmoid	10	–	0.5	–	–	94
Sigmoid	10	–	0.2	–	–	93

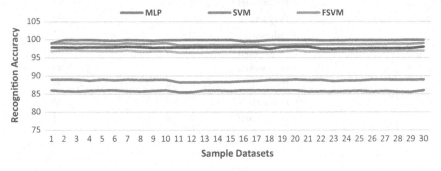

Comparative Performance of Soft Computing vs Traditional Techniques: French Language

Fig. 5.13 The comparative performance of soft computing versus traditional techniques for French language

be categorized as ascenders (characters such as ç, Ç etc.) and descenders (characters such as þ, Á etc.) as illustrated in Sect. 5.8.1. HFBRNN shows better results than the traditional methods as well RFMLP [7]. The testing algorithm is applied to all standard cases of characters [26] in various samples. An accuracy of around 99.9% has been achieved in all the cases.

5.9   Further Discussions

All the feature based classification based methods used in this chapter give better results than the traditional approaches [6]. On comparing with other algorithms, it is observed that RFMLP work with high accuracy in almost all cases

Table 5.9 The experimental results for a subset of the French characters using HFBRNN

Characters	Successful recognition (%)	Unsuccessful recognition (%)	No recognition (%)
á	99.9	0.1	0
â	99.9	0.1	0
ã	99.9	0.1	0
ä	99.9	0.1	0
å	99.9	0.1	0
ç	99.9	0.1	0
è	99.9	0.1	0
é	99.9	0.1	0
ê	99.9	0.1	0
ë	99.9	0.1	0
ì	99.9	0.1	0
í	99.9	0.1	0
î	99.9	0.1	0
ï	99.9	0.1	0
ò	99.9	0.1	0
þ	99.9	0.1	0
û	99.9	0.1	0
ú	99.9	0.1	0
ù	99.9	0.1	0
ó	99.9	0.1	0
ô	99.9	0.1	0
õ	99.9	0.1	0
ö	99.9	0.1	0
ñ	99.9	0.1	0
ý	99.9	0.1	0
ÿ	99.9	0.1	0
Á	99.9	0.1	0
Â	99.9	0.1	0
Ã	99.9	0.1	0
Ä	99.9	0.1	0
Å	99.9	0.1	0
Ç	99.9	0.1	0
È	99.9	0.1	0
É	99.9	0.1	0
Ê	99.9	0.1	0
Ë	99.9	0.1	0
Ì	99.9	0.1	0
Í	99.9	0.1	0
Î	99.9	0.1	0
Ï	99.9	0.1	0

(continued)

Table 5.9   (continued)

Characters	Successful recognition (%)	Unsuccessful recognition (%)	No recognition (%)
Ò	99.9	0.1	0
Ó	99.9	0.1	0
Ô	99.9	0.1	0
Õ	99.9	0.1	0
Ö	99.9	0.1	0
Ù	99.9	0.1	0
Ú	99.9	0.1	0
Ü	99.9	0.1	0
Ý	99.9	0.1	0
þ	99.9	0.1	0
Ð	99.9	0.1	0
Ñ	99.9	0.1	0

which include intersections of loop and instances of multiple crossings. This algorithm focussed on the processing of various asymmetries in the French characters. RFMLP function consider the conditions: (a) the validity of the algorithms for non-cursive French alphabets (b) the requirement of the algorithm that height of both upper and lower case characters to be proportionally same and (c) for extremely illegible handwriting the accuracy achieved by the algorithm is very less.

FSVM and FRSVM also give superior performance compared to traditional SVM [6]. FSVM and FRSVM achieve a precision rate up to 97% in case of training with sets corresponding to either small or capital letters and up to 98% in case of training with sets corresponding to both small and capital letters respectively. Thus the system achieved its goal through the recognition of characters from an image. The future research direction entails in expanding the system through addition of techniques which determine automatically the optimal parameters of kernel functions. Further, any version of SVM can be used to better perform the feature based classification task. We are also experimenting with other robust versions of SVM which will improve the overall recognition accuracy of the French characters.

HFBRNN produces the best results in terms of accuracy for all cases including loop intersections and multiple crossing instances. This algorithm also focussed on the processing of various asymmetries in the French characters as RFMLP. HFBRNN achieves high successful recognition rate of about 99.9% for all the French characters.

Finally we conclude the chapter with a note that we are exploring further results on IRESTE IRONFF dataset using other soft computing techniques such as fuzzy markov random fields and rough version of hierarchical bidirectional recurrent neural networks.

References

1. Bunke, H., Wang, P. S. P. (Editors), Handbook of Character Recognition and Document Image Analysis, World Scientific, 1997.
2. Chaudhuri, A., Ghosh, S. K., Sentiment Analysis of Customer Reviews Using Robust Hierarchical Bidirectional Recurrent Neural Network, Book Chapter: Artificial Intelligence Perspectives in Intelligent Systems, Radek Silhavy, Roman Senkerik, Zuzana Kominkova Oplatkova, Petr Silhavy, Zdenka Prokopova, (Editors), Advances in Intelligent Systems and Computing, Springer International Publishing, Switzerland, Volume 464, pp 249–261, 2016.
3. Chaudhuri, A., Fuzzy Rough Support Vector Machine for Data Classification, International Journal of Fuzzy System Applications, 5(2), pp 26–53, 2016.
4. Chaudhuri, A., Modified Fuzzy Support Vector Machine for Credit Approval Classification, AI Communications, 27(2), pp 189–211, 2014.
5. Chaudhuri, A., De, Fuzzy Support Vector Machine for Bankruptcy Prediction, Applied Soft Computing, 11(2), pp 2472–2486, 2011.
6. Chaudhuri, A., Applications of Support Vector Machines in Engineering and Science, Technical Report, Birla Institute of Technology Mesra, Patna Campus, India, 2011.
7. Chaudhuri, A., Some Experiments on Optical Character Recognition Systems for different Languages using Soft Computing Techniques, Technical Report, Birla Institute of Technology Mesra, Patna Campus, India, 2010.
8. Chaudhuri, A., De, K., Job Scheduling using Rough Fuzzy Multi-Layer Perception Networks, Journal of Artificial Intelligence: Theory and Applications, 1(1), pp 4–19, 2010.
9. Chaudhuri, A., De, K., Chatterjee, D., Discovering Stock Price Prediction Rules of Bombay Stock Exchange using Rough Fuzzy Multi-Layer Perception Networks, Book Chapter: Forecasting Financial Markets in India, Rudra P. Pradhan, Indian Institute of Technology Kharagpur, (Editor), Allied Publishers, India, pp 69–96, 2009.
10. Cheriet, M., Kharma, N., Liu, C. L., Suen, C. Y., Character Recognition Systems: A Guide for Students and Practitioners, John Wiley and Sons, 2007.
11. De, R. K., Basak, J., Pal, S. K., Neuro-Fuzzy Feature Evaluation with Theoretical Analysis, Neural Networks, 12(10), pp 1429–1455, 1999.
12. De, R. K., Pal, N. R., Pal, S. K., Feature Analysis: Neural Network and Fuzzy Set Theoretic Approaches, Pattern Recognition, 30(10), pp 1579–1590, 1997.
13. Gonzalez, R. C., Woods, R. E., Digital Image Processing, 3rd Edition, Pearson, 2013.
14. Jain, A. K., Duin, R. P. W., Mao, J., Statistical Pattern Recognition: A Review, IEEE Transactions on Pattern Analysis and Machine Intelligence, 22(1), pp 4–37, 2000.
15. Jain, A. K., Fundamentals of Digital Image Processing, Prentice Hall, India, 2006.
16. Pal, S. K., Mitra, S., Mitra, P., Rough-Fuzzy Multilayer Perception: Modular Evolution, Rule Generation and Evaluation, IEEE Transactions on Knowledge and Data Engineering, 15(1), pp 14–25, 2003.
17. Russ, J. C., The Image Processing Handbook, CRC Press, 6th Edition, 2011.
18. Schantz, H. F., The History of OCR, Recognition Technology Users Association, Manchester Centre, VT, 1982.
19. Taghva, K., Borsack, J., Condit, A., Effects of OCR Errors on Ranking and Feedback using the Vector Space Model, Information Processing and Management, 32(3), pp 317–327, 1996.
20. Taghva, K., Borsack, J., Condit, A., Evaluation of Model Based Retrieval Effectiveness with OCR Text, ACM Transactions on Information Systems, 14(1), pp 64–93, 1996.

21. Taghva, K., Borsack, J., Condit, A., Erva, S., The Effects of Noisy Data on Text Retrieval, Journal of American Society for Information Science, 45 (1), pp 50–58, 1994.
22. Young, T. Y., Fu, K. S., Handbook of Pattern Recognition and Image Processing, Academic Press, 1986.
23. Zadeh, L. A., Fuzzy Sets, Information and Control, 8(3), pp 338–353, 1965.
24. Zimmermann, H. J., Fuzzy Set Theory and its Applications, 4th Edition, Kluwer Academic Publishers, Boston, 2001.
25. https://en.wikipedia.org/wiki/French_language.
26. http://www.infres.enst.fr/~elc/GRCE/news/IRONOFF.doc.
27. https://www.abbyy.com/finereader/.
28. https://www.csie.ntu.edu.tw/~cjlin/libsvm/.

Chapter 6
Optical Character Recognition Systems for German Language

Abstract The optical character recognition (OCR) systems for German language were the most primitive ones and occupy a significant place in pattern recognition. The German language OCR systems have been used successfully in a wide array of commercial applications. The different challenges involved in the OCR systems for German language is investigated in this chapter. The pre-processing activities such as text region extraction, skew detection and correction, binarization, noise removal, character segmentation and thinning are performed on the datasets considered. The feature extraction is performed through fuzzy Genetic Algorithms (GA). The feature based classification is performed through important soft computing techniques viz rough fuzzy multilayer perception (RFMLP), fuzzy support vector machine (FSVM), fuzzy rough support vector machine (FRSVM) and hierarchical fuzzy bidirectional recurrent neural networks (HFBRNN). The superiority of soft computing techniques is demonstrated through the experimental results.

Keywords German language OCR · RFMLP · FSVM · FRSVM · HFBRNN

6.1 Introduction

The optical character recognition (OCR) for German language [7, 18] places itself in an important category of systems in pattern recognition. This has led to the development of several OCR systems for the German language [1]. German is the most widely spoken in Central Europe. The German language OCR systems have been used successfully in a wide array of commercial products [7]. The character recognition of German language has a high potential in data and word processing as the English and French languages. Some commonly used applications of the OCR systems of German language [7] are automated postal address and ZIP code reading, data acquisition in bank checks, processing of archived institutional records etc.

© Springer International Publishing AG 2017   137
A. Chaudhuri et al., *Optical Character Recognition Systems for Different Languages with Soft Computing*, Studies in Fuzziness and Soft Computing 352,
DOI 10.1007/978-3-319-50252-6_6

The standardization of OCR character set for German language was provided through ISO/IEC 646 [25]. ISO/IEC 646 is a joint ISO and IEC series of standards for 7-bit coded character set for information interchange and development in cooperation with ASCII since 1964. ISO/IEC 646 was also ratified by ECMA as ECMA-6 first published in 1965. The characters in the ISO/IEC 646 basic character set are invariant characters. This was following the similar specifications as those used by the ISO basic Latin alphabets. Since the universal acceptance of the 8-bit byte did not exist then, the German characters had to be made to fit within the constraints of 7 bits. This means that some characters that appear in ASCII do not appear in other national variants of ISO 646. In the recent years, the OCR for German language has gained a considerable importance as the need for converting the scanned images into computer recognizable formats such as text documents has variety of applications. The German language based OCR systems is thus one of the most fascinating and challenging areas of pattern recognition with various practical applications [7].

The OCR process for any language involves extraction of defined characteristics called features to classify an unknown character into one of the known classes [1, 6, 7, 10] to a user defined accuracy level. As such any good OCR system is best defined in terms of the success of feature extraction and classification tasks. The same is true for the German language. The process becomes tedious in case the language has dependencies where some characters look identical. Thus the classification becomes a big challenge.

In this chapter we start the investigation of OCR systems considering the different challenges involved in German language. The different pre-processing activities such as text region extraction, skew detection and correction, binarization, noise removal, character segmentation and thinning are performed on the considered datasets [26]. The feature extraction is performed through fuzzy genetic algorithms. The feature based classification is performed through important soft computing techniques viz rough fuzzy multilayer perceptron (RFMLP) [8, 9, 16] two support vector machine (SVM) based methods such as fuzzy support vector machine (FSVM) [4, 5] and fuzzy rough support vector machine (FRSVM) [3] and hierarchical fuzzy bidirectional recurrent neural networks (HFBRNN) [2]. The experimental results demonstrate the superiority of soft computing techniques over the traditional methods.

This chapter is structured as follows. In Sect. 6.2 a brief discussion about the German language script and datasets used for experiments are presented. The different challenges of OCR for German language are highlighted in Sect. 6.3. The next section illustrates the data acquisition. In Sect. 6.5 different pre-processing activities on the datasets such as text region extraction, skew detection and correction, binarization, noise removal, character segmentation and thinning are presented. This is followed by a discussion of feature extraction on German language dataset in Sect. 6.6. The Sect. 6.7 explains the state of art of OCR for German language in terms of feature based classification methods. The corresponding experimental results are given in Sect. 6.8. Finally Sect. 6.9 concludes the chapter with some discussions and future research directions.

6.2   German Language Script and Experimental Dataset

In this section we present brief information about the German language script and the dataset used for experiments. German is one of the major languages of the world [25] and the first language of about 100 million people worldwide. It is the most widely spoken native language in the European Union. German is the most widely spoken and language in Germany, Austria, Switzerland, South Tyrol, Liechtenstein, Luxembourg and Belgium. It is similar to other West Germanic languages such as Afrikaans, Dutch and English. German is the third most widely taught foreign language in both the United States (after Spanish and French) and the European Union (after English and French) at lower secondary level. It is the second most commonly used scientific language as well as the third most widely used language on websites (after English and Russian). The German language is ranked fifth in terms of annual publication of new books with one tenth of all books in the world being published in German.

The German language derives most of its vocabulary from the Germanic branch of the Indo-European language family [25]. A subset of German vocabulary are derived from Latin and Greek and the rest are borrowed from French and English. With slightly different standardized variants like German, Austrian and Swiss Standard German, German is a pluricentric language. The German language is also notable for its broad spectrum of dialects with many unique varieties existing in Europe and also other parts of the world.

The history of the German language begins with the High German consonant shift during the migration period which separated Old High German dialects from Old Saxon. The earliest evidence of Old High German is from scattered Elder Futhark inscriptions especially in Alemannic from the sixth century AD; the earliest glosses (*Abrogans*) date to the eighth century and the oldest coherent texts (*Hildebrandslied*, *Muspilli* and Merseburg Incantations) to the ninth century. The Old Saxon belonged to the North Sea Germanic cultural sphere and Lower Saxony was under German rather than Anglo-Frisian influence during the existence of the Holy Roman Empire.

The German language dataset used for performing OCR experiments is the InftyCDB-2 dataset and is adapted from [26]. InftyCDB-2 database contains German text which is used here to for training and testing. The database contains unconstrained handwritten text which are scanned at a resolution of 300 dpi and saved as PNG images with 256 gray levels. The Fig. 6.1 shows a sample snapshot from the database. InftyCDB-2 database contains 77,820 German characters where few hundred writers contributed samples of their handwriting.

Further details are available at [26].

German Alphabet					
Aa ah	Ää ah Umlaut	Bb beh	ß ess-testt	Cc tseh	Dd deh
Ee eh	Ff eff	Gg geh	Hh ha	Ii ee	Jj yot
Kk kah	Ll ell	Mm emm	Nn enn	Oo oh	Öö oh umlaut
Pp peh	Qq kuh	Rr err	Ss ess	Tt teh	Uu uh
Üü uh Umlaut	Vv fow	Ww veh	Xx iks	Yy upsilon	Zz tsett

Fig. 6.1   A sample text snapshot from InftyCDB-2 database

6.3 Challenges of Optical Character Recognition Systems for German Language

After English and French languages the OCR for German language has become one of the most successful applications of technology in pattern recognition and artificial intelligence. The OCR for German language has been the topic of active research since past few decades [7]. The most commercially available OCR system for German language is i2OCR [27] which supports major language formats and multi column document analysis. Considering the important aspects of versatility, robustness and efficiency, the commercial OCR systems are generally divided into four generations [7] as highlighted in Chap. 2. It is to be noted that this categorization refers also to the OCRs of German language.

Despite decades of research and existence of established commercial OCR products based on German language, the output from such OCR processes often contains errors. The more highly degraded is input, the greater is error rate. Since inputs form the first stage in a pipeline where later stages are designed to support sophisticated information extraction and exploitation applications, it is important to understand the effects of recognition errors on downstream analysis routines. Few questions are required to be addressed in this direction. They are as follows:

(a)   Are all recognition errors equal in impact or some are worse than others?

(b)   Can the performance of each stage be optimized in isolation or the end-to-end system should be considered?

(c)   In balancing the trade-off between the risk of over and under segmenting characters during OCR where should the line be drawn to maximize overall performance?

The answers to these questions often influence the way OCR systems for German language are designed and build for analysis [7].

The German language OCR system converts numerous published books in German language into editable computer text files. The latest research in this area has grown to incorporate some new methodologies to overcome the complexity of German writing style. All these algorithms have still not been tested for complete characters of German alphabet. Hence, there is a quest for developing an OCR system which handles all classes of German text and identify characters among these classes increasing versatility, robustness and efficiency in commercial OCR systems. The recognition of printed German characters is itself a challenging problem since there is a variation of the same character due to change of fonts or introduction of different types of noises. There may be noise pixels that are introduced due to scanning of the image. A significant amount of research has been done towards text data processing in German language from noisy sources [7]. The majority of the work has focused predominately on errors that arise during speech recognition systems [21, 22]. Several research papers have appeared which examines the noise problem from variety of perspectives with most emphasizing issues that are inherent in written and spoken German language [7]. However, there has been less work concentrating on noise induced by OCR. Some earlier works by [19] show that moderate error rates have little impact on effectiveness of traditional information retrieval measures. However, this conclusion is tied to certain assumptions about information retrieval through bag of words, OCR error rate which may not be too high and length of documents which may not be too short. Some other notable research works in this direction are given in [20, 21]. All these works try to address some significant issues involved in OCR systems for German language such as error correction, performance evaluation etc. involving flexible and rigorous mathematical treatment [7]. Besides this any German character can be represented in variety of fonts and sizes as per the needs and requirements of application. Further the character with same font and size may also be bold face character as well as normal one [9]. Thus the width of stroke also significantly affects recognition process. Therefore, a good character recognition approach for German language [7, 15, 17, 22]:

(a)   Must eliminate noise after reading binary image data
(b)   Smooth image for better recognition
(c)   Extract features efficiently
(d)   Train the system and
(e)   Classify patterns accordingly.

6.4 Data Acquisition

The progress in automatic character recognition systems in German language is motivated from two categories according to the mode of data acquisition which can be either online or offline character recognition systems [7]. Following the lines of English language, the data acquisition of German language can be either online or offline character recognition systems. The offline character recognition captures data from paper through optical scanners or cameras whereas online recognition systems utilize digitizers which directly capture writing with the order of strokes, speed, pen-up and pen-down information. As such the scope of this text is restricted to OCR systems, we confine our discussion to offline character recognition [7] for German language. The German language datasets used in this research is mentioned in Sect. 6.2.

6.5 Data Pre-processing

Once the data has been acquired properly we proceed to pre-process the data. In pre-processing stage [13, 14, 17] a series of operations are performed here which includes text region extraction, skew detection and correction, binarization, noise removal, character segmentation and thinning or skeletonization. The main objective of pre-processing is to organize information so that the subsequent character recognition task becomes simpler. It essentially enhances the image rendering it suitable for segmentation.

6.5.1 Text Region Extraction

The text region extraction is used here as the first step in character recognition process. The input image $I_{P \times Q}$ is first partitioned into m number of blocks $B_i; i = 1, 2, \ldots \ldots, m$ such that:

$$B_i \cap B_j = \emptyset \tag{6.1}$$

$$I_{P \times Q} = \cup_{i=1}^{m} B_i \tag{6.2}$$

A block B_i is a set of pixels represented as $B_i = [f(x, y)]_{H \times W}$ where H and W are the height and the width of the block respectively. Each individual block B_i is classified as either information block or background block based on the intensity variation within it. After removal of background blocks adjacent or contiguous information blocks constitute isolated components called as regions $R_i; i = 1, 2, \ldots, n$ such that:

$$R_i \cap R_j = \emptyset \, \forall \, \text{but} \, \cup_{i=1}^{n} R_i \neq I_{P \times Q} \tag{6.3}$$

This is because some background blocks have been removed. The area of a region is always a multiple of the area of the blocks. These regions are then classified as

Fig. 6.2 The camera captured image and the text regions extracted from it

text region or non-text region using various characteristics features of textual and non-textual regions such as dimensions, aspect ratio, information pixel density, region area, coverage ratio, histogram, etc. A detail description of this technique has been presented in [7]. The Fig. 6.2 shows a camera captured image and the text regions extracted from it.

6.5.2 Skew Detection and Correction

When a text document is fed into scanner either mechanically or manually a few degrees of tilt or skew is unavoidable. In skew angle the text lines in digital image make angle with horizontal direction. A number of methods are available in literature for identifying image skew angles [1]. They are basically categorized on the basis of projection profile analysis, nearest neighbor clustering, Hough transform, cross correlation and morphological transforms. The camera captured images very often suffer from skew and perspective distortion [15]. They occur due to non-parallel axes or planes at the time of capturing the image. The acquired image does not become uniformly skewed mainly due to perspective distortion. The skewness of different portions of the image may vary between $+\alpha$ to $-\beta$ degrees where both α and β are positive numbers. Hence, the image cannot be deskewed at a single pass. On the other hand the effect of perspective distortion is distributed throughout the image. Its effect is hardly visible within a small region for example, the area of a character of the image. At the same time the image segmentation module generates only a few text regions. These text regions are deskewed using a computationally efficient and fast skew correction technique presented in [7].

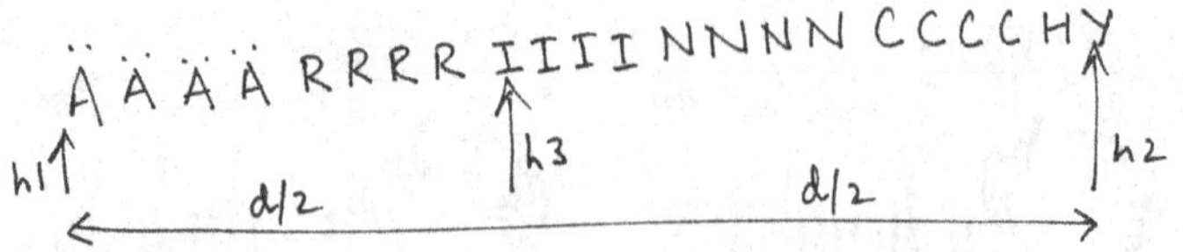

Fig. 6.3 The calculation of skew angle from bottom profile of a text region

Every text region basically has two types of pixels viz dark and gray. The dark pixels constitute the texts and the gray pixels are background around the texts. For the four sides of virtual bounding rectangle of a text region, there are four sets of values known as profiles. If the length and breadth of the bounding rectangle are M and N respectively, then two profiles will have M values each and the other two will have N values each. These values are the distances in terms of pixel from a side to the first gray or black pixel of the text region. Among these four profiles, the one which is from the bottom side of the text region is taken into consideration for estimating skew angle as shown in Fig. 6.3. This bottom profile is denoted as $\{h_i; i = 1, 2, \ldots, M\}$.

The mean $\mu = \frac{1}{M} \sum_{i=1}^{M} h_i$ and the first order moment $\tau = \frac{1}{M} \sum_{i=1}^{M} |\mu - h_i|$ values are calculated. Then, the profile size is reduced by excluding some h_i values that are not within the range $\mu \pm \tau$. The central idea behind this exclusion is that these elements hardly contribute to the actual skew of the text region. Now from the remaining profile, the elements viz leftmost h_1, rightmost h_2 and the middle one h_3 are chosen. The final skew angle is computed by averaging the three skew angles obtained from three pairs h_1–h_3, h_3–h_2 and h_1–h_2. Once the skew angle for a text region is estimated it is rotated by the same angle.

6.5.3 Binarization

Binarization [13, 17] is the next step in character recognition process. A large number of binarization techniques are available in the literature [7] each of which is appropriate to particular image types. Its goal is to reduce the amount of information present in the image and keep only the relevant information. Generally the binarization techniques of gray scale images are classified into two categories viz overall threshold where single threshold is used in the entire image to form two classes (text and background) and local threshold where values of thresholds are determined locally (pixel-by-pixel or region-by-region). Here the skew corrected text region is binaries using an efficient binarization technique [22]. The algorithm is given below:

Binarization Algorithm:
begin
  for all pixels in (x, y) in *Text Region*
    if intensity $(x, y) < (G_{max} + G_{min})/2$
      then mark (x, y) as foreground
      else if number of foreground neighbors > 4
      then mark (x, y) as foreground
      else mark (x, y) as background
      end if
    end if
  end for
end

This is an improved version of bernsen's binarization method [7]. The arithmetic mean of maximum G_{max} and minimum G_{min} gray levels around a pixel is taken as the threshold for binarizing the pixel. In the present algorithm the eight immediate neighbors around the pixel subject to binarization are also taken as deciding factors for binarization. This approach is especially useful to connect the disconnected foreground pixels of a character.

6.5.4 Noise Removal

The scanned text documents often contain noise that arises due to printer, scanner, print quality, document age etc. Therefore, it is necessary to filter noise [13] before the image is processed. Here a low-pass filter is used to process the image [15] which is used for later processing. The main objective in the design of a noise filter is that it should remove as much noise as possible while retaining the entire signal [17].

6.5.5 Character Segmentation

Once the text image is skew corrected, binarized and noise removed, the actual text content is extracted. This process leads to character segmentation [13]. The commonly used segmentation algorithms in this direction are connected component labeling, x-y tree decomposition, run length smearing and Hough transform [7]. After binarizing a noise free text region, the horizontal histogram profile $\{f_i; i = 1, 2, \ldots, H_R\}$ of the region as shown in Fig. 6.4a, b is analyzed for segmenting the region into text lines. Here f_i denotes the number of black pixel along ith of the text region and hidden region denotes the height of the deskewed text region. Firstly, all possible line segments are determined by thresholding the

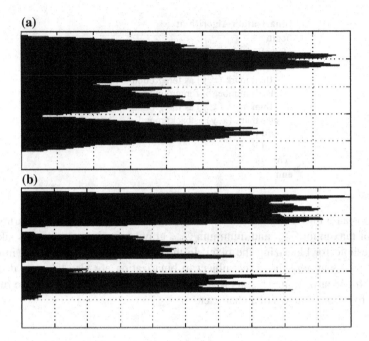

Fig. 6.4 a The horizontal histogram of text regions for their segmentation (skewed text region and its horizontal histogram). **b** The horizontal histogram of text regions for their segmentation (skew corrected text region and its horizontal histogram)

profile values. The threshold is chosen so as to allow over segmentation. Text line boundaries are referred by the values of i for which the value of f_i is less than the threshold. Thus, n such segments represent $n - 1$ text lines. After that the inter segment distances are analyzed and some segments are rejected based on the idea that the distance between two lines in terms of pixels will not be too small and the inter-segment distances are likely to become equal. A detail description of the method is given in [15]. Using vertical histogram profile of each individual text lines, words and characters are segmented. Sample segmented characters have been shown in the Fig. 6.5a, b.

6.5.6 Thinning

The character segmentation process is followed by thinning or skeletonization. In thinning one-pixel-width representation or skeleton of an object is obtained by preserving connectedness of the object and its end points [17]. The thinning process reduces image components to their essential information so that further analysis and recognition are facilitated. For instance, an alphabet can be handwritten with different pens giving different stroke thicknesses but information presented

Fig. 6.5 **a** Skew correction and segmentation of text regions (an extracted text region). **b** Skew correction and segmentation of text regions (characters segmented from de-skewed text region)

(a)

(b)

is same. This enables easier subsequent detection of pertinent features. As an illustration consider letter **ä** shown in Fig. 6.6 before and after thinning. A number of thinning algorithms have been used in the past with considerable success. The most common algorithm used is the classical hilditch algorithm and its variants [15]. Here hilditch algorithm is used for thinning [7]. For recognizing large graphical objects with filled regions which are often found in logos, region boundary detection is useful but for small regions corresponding to individual characters neither thinning nor boundary detection is performed. Rather entire pixel array representing the region is forwarded to subsequent stage of analysis.

Fig. 6.6 An image before
and after thinning

6.6 Feature Selection Through Genetic Algorithms

As mentioned in Chap. 4 the heart of any OCR system is the formation of feature
vector used in recognition stage. This fact is also valid for German language OCR
system. This phase extracts the features from segmented areas of image containing characters to be recognized that distinguishes an area corresponding to a letter from an area corresponding to other letters. The feature extraction phase can
thus be visualised as finding a set of parameters or features that define the shape
of character as precise and unique. In Chap. 3 the term feature extraction is often
used synonymously by feature selection which refers to algorithms that select the
best subset of input feature set. These methods create new features based on transformations or combination of original features [7]. The features selected help in
discriminating the characters. Achieving high recognition performance is attributed towards the selection of appropriate feature extraction methods. A large number of OCR based feature extraction methods are available in literature [7] except
that the selected method depends on the application concerned. There is no universally accepted set of feature vectors in OCR. The features that capture topological and geometrical shape information are the most desired ones. The features that
capture spatial distribution of black (text) pixels are also very important [13]. The
genetic algorithm based feature extraction approach based on multilayer perceptron (MLP) is successfully applied for the OCR of German language [7].

A number of neural network and fuzzy set theoretic approaches [23, 24] have
been proposed for feature analysis in recent past [7]. A feature quality index (FQI)
measure for ranking of features has also been suggested [12]. The feature ranking
process is based on influence of feature on MLP output. It is related to the importance of feature in discriminating among classes. The impact of qth feature on MLP
output out of a total of p features is measured by setting feature value to zero for
each input pattern $x_i, i = 1, \ldots, n$. FQI is defined as the deviation of MLP output
with qth feature value set to zero from output with all features present such that:

$$FQI_q = \frac{1}{n} \sum\nolimits_{i=1}^{n} \left\| OV_i - OV_i^{(q)} \right\|^2 \tag{6.4}$$

In Eq. (6.4), OV_i and $OV_i^{(q)}$ are output vectors with all p features present and with
qth feature set to zero. The features are ranked according to their importance as
$q_1, \ldots, q_p$ if $FQI_{q_1} > \cdots > FQI_{q_p}$. In order to select best p' features from the set

of p features, $\binom{p}{p'}$ possible subsets are tested one at a time. The quality index $FQI_k^{(p')}$ of kth subset S_k is measured as:

$$FQI_k^{(p')} = \frac{1}{n}\sum_{i=1}^{n}\left\|OV_i - OV_i^k\right\|^2 \tag{6.5}$$

In Eq. (6.5), OV_i^k is MLP output vector with x_i^k as input where x_i^k is derived from x_i as:

$$x_{ij}^k = \begin{cases} 0 & \text{if } j \in S_k \\ x_{ij} & \text{ow} \end{cases} \tag{6.6}$$

A subset S_j is selected as optimal set of features if $FQI_j^{(p')} \geq FQI_k^{(p')} \forall k, k \neq j$. An important observation here is that value of p' should be predetermined and $\binom{p}{p'}$ number of possible choices are verified to arrive at the best feature set. It is evident that no apriori knowledge is usually available to select the value p' and an exhaustive search is to be made for all values p' with $p' = 1, \ldots, p$. The number of possible trials is $(2^p - 1)$ which is prohibitively large for high values of p. To overcome drawbacks of above method, best feature set is selected by using of genetic algorithms [7]. Let us consider mask vector M where $M_i \in \{0, 1\}; i = 1, \ldots, p$ and each feature element $q_i, i = 1, \ldots, n$ is multiplied by corresponding mask vector element before reaching MLP input such that $I_i = q_i M_i$. MLP inputs are then written as:

$$I_i = \begin{cases} 0 & \text{if } M_i = 0 \\ q_i & \text{ow} \end{cases} \tag{6.7}$$

Thus, a particular feature q_i reaches MLP if corresponding mask element is one. To find sensitivity of a particular feature q_j, mask bit M_j is set to zero. With respect to the above discussions when kth subset of feature set $\{q_1, \ldots, q_p\}$ is selected, all corresponding mask bits are set to zero and rest are set to one. When feature set multiplied by these mask bits reaches MLP, the effect of setting features of subset S_k to zero is obtained. Then value of FQI_k is calculated. It is to be noted that kth subset thus chosen may contain any number of feature elements and not pre-specified p' number of elements. Starting with an initial population of strings representing mask vectors, genetic algorithm is used with reproduction, crossover and mutation operators to determine best value of objective function. The objective function is FQI value of feature set S_k selected with mask bits set to zero for specific features and is given by:

$$FQI_k = \frac{1}{n}\sum_{i=1}^{n}\left\|OV_i - OV_i^k\right\|^2 \tag{6.8}$$

Fig. 6.7 The feature
selection process using
genetic algorithm

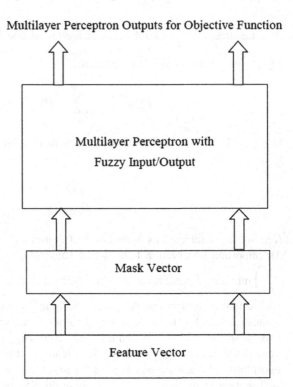

In this process, both the problems of predetermining value of p' and searching through $\binom{p}{p'}$ possible combinations for each value of p'. In genetic algorithm implementation, the process is started with 20 features generated from fuzzy Hough transform such that the number of elements in mask vector is also 20. After running genetic algorithm for sufficiently large number of generations, mask string with best objective function value is determined. The feature elements corresponding to mask bits zero are chosen as selected set of features. The parameters for genetic algorithms are determined in terms of chromosome length, population size, mutation probability and crossover probability. MLP is next trained with only selected feature set for classification. The number of features varied when required. The genetic algorithm based feature selection method is shown in Fig. 6.7.

6.7 Feature Based Classification: Sate of Art

After extracting the essential features from the pre-processed character image the concentration pointer turns on the feature based classification methods for OCR of German language. Considering the wide array of feature based classification methods for OCR, these methods are generally grouped into following four broad categories [11, 12, 20]:

(a)   Statistical methods
(b)   ANN based methods
(c)   Kernel based methods
(d)   Multiple classifier combination methods

We discuss here the work done on German characters using some of the above-mentioned classification methods. All the classification methods used here are soft computing based techniques. The next three subsections highlight the feature based classification based on RFMLP [8, 9, 16], FSVM [4, 5], FRSVM [3] and HFBRNN [2] techniques. These methods are already discussed in Chap. 3. For further details interested readers can refer [6, 7].

6.7.1 Feature Based Classification Through Rough Fuzzy Multilayer Perceptron

The German language characters are recognized here through RFMLP [8, 9, 16] which is discussed in Chap. 3. For the German language RFMLP has been used with considerable success [7]. The diagonal feature extraction scheme is used for drawing of features from the characters. For classification stage we use these features extracted. RFMLP used here is a feed forward network with back propagation ANN having four hidden layers. The architecture used is 70–100–100–100–100–100–52 [7] for classification. The four hidden layers and output layer uses tangent sigmoid as the activation function. The feature vector is again denoted by $F = (f_1, \ldots, f_m)$ where m denotes number of training images and each f has a length of 70 which represent the number of input nodes. The 59 neurons in the output layer correspond to the 30 lowercase and 29 uppercase German alphabets [26]. The network training parameters are briefly summarized as [7]:

(a)   Input nodes: 70
(b)   Hidden nodes: 100 at each layer
(c)   Output nodes: 59
(d)   Training algorithm: Scaled conjugate gradient backpropagation
(e)   Performance function: Mean Square Error
(f)   Training goal achieved: 0.000 002
(g)   Training epochs: 4000.

6.7.2   Feature Based Classification Through Fuzzy and Fuzzy Rough Support Vector Machines

In similar lines to the English language character recognition through FSVM and FRSVM in Sect. 4.7.3, FSVM [4, 5] and FRSVM [3] discussed in Chap. 3 is used here to recognize the German language characters. Over the past years SVM based methods have shown a considerable success in feature based classification [6]. With this motivation FSVM and FRSVM are used here for feature classification task. Both FSVM and FRSVM offer the possibility of selecting different types of kernel functions such as sigmoid, RBF, linear functions and determining the best possible values for these kernel parameters [4–6]. After selecting the kernel type and its parameters, FSVM and FRSVM are trained with the set of features obtained from other phases. Once the training gets over, FSVM and FRSVM are used to classify new character sets. The implementation is achieved through LibSVM library [28].

6.7.3   Feature Based Classification Through Hierarchical Fuzzy Bidirectional Recurrent Neural Networks

After RFMLP, FSVM and FRSVM, the German language characters are recognized through HFBRNN [2] which is discussed in Chap. 3. For the German language HFBRNN has been used with considerable success [7]. HBRNN is a deep learning based technique [2]. In the recent past deep learning based techniques have shown a considerable success in feature based classification [2]. With this motivation HBRNN is used here for feature classification task and takes full advantage of deep recurrent neural network (RNN) towards modeling long-term information of data sequences. The recognition of characters done by HFBRNN at review level. The performance of HFBRNN is improved by fine tuning parameters of the network in a hierarchical fashion. The motivation is obtained from long short term memory (LSTM) and bidirectional LSTM (BLSTM) [2]. The evaluation is done on different types of highly biased character data. The implementation of HBRNN is performed in MATLAB [7].

6.8   Experimental Results

In this section, the experimental results for soft computing tools viz RFMLP, FSVM, FRSVM and HFBRNN on the German language dataset [26] highlighted in Sect. 6.2 are presented. The prima face is to select the best OCR system for German language [7].

6.8.1 Rough Fuzzy Multilayer Perceptron

The experimental results for a subset of the German characters both uppercase and lowercase using RFMLP are shown in the Table 6.1. Like the English characters, the German characters can be categorized as ascenders (characters such as b, d, h etc.) and descenders (characters such as P, p, q, y etc.). RFMLP shows better results than the traditional methods [7]. The testing algorithm is applied to all standard cases of characters [26] in various samples. An accuracy of around 99%

Table 6.1 The experimental results for a subset of the German characters using RFMLP

Characters	Successful recognition (%)	Unsuccessful recognition (%)	No recognition (%)
a	99	1	0
ä	99	1	0
b	99	1	0
β	99	1	0
c	99	1	0
d	99	1	0
e	98	1	1
f	99	1	0
g	99	1	0
h	99	1	0
i	99	1	0
j	99	1	0
k	9S	1	1
1	98	1	1
m	99	1	0
n	99	1	0
o	99	1	0
ö	99	1	0
p	99	1	0
q	99	1	0
r	99	1	0
s	99	1	0
t	98	1	1
u	98	1	1
ü	99	1	0
v	99	1	0
w	99	1	0
x	99	1	0
y	99	1	0
z	99	1	0

(continued)

Table 6.1 (continued)

Characters	Successful recognition (%)	Unsuccessful recognition (%)	No recognition (%)
A	99	1	0
Ä	99	1	0
B	99	1	0
C	99	1	0
D	99	1	0
E	99	1	0
F	99	1	0
G	99	1	0
H	98	1	1
I	98	1	1
J	99	1	0
K	99	1	0
L	99	1	0
M	99	1	0
N	99	1	0
O	99	1	0
Ö	99	1	0
P	99	1	0
Q	98	1	1
R	98	1	1
S	99	1	0
T	99	1	0
U	99	1	0
Ü	99	1	0
V	99	1	0
W	99	1	0
X	99	1	0
Y	99	1	0
Z	99	1	0

has been achieved in all the cases. However, after successful training and testing of algorithm the following flaws are encountered [7] which are identical to those encountered in English language:

(a) There may an instance when there is an occurrence of large and disproportionate symmetry in both ascenders as well as descenders as shown in Fig. 6.8a, b.

(b) There may be certain slants in certain characters which results in incorrect detection of characters as shown in Fig. 6.9.

(c) Sometimes one side of the image may have less white space and more pixel concentration whereas other side may have more white space and less pixel concentration. Due to this some characters are wrongly detected as shown in Fig. 6.10.

Fig. 6.8 The disproportionate symmetry in characters 'p' and 't'

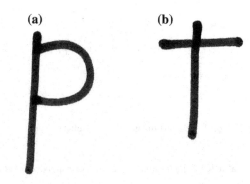

(a)　　　　**(b)**

Fig. 6.9 The character 't' with slant

Fig. 6.10 The uneven pixel combination for character '*k*' (region-i on the *left image* has less white space than the *right image*)

6.8.2 Fuzzy and Fuzzy Rough Support Vector Machines

FSVM and FRSVM show promising results for feature classification task through different types of kernel functions by selecting the best possible values for kernel parameters [3–5]. For testing accuracy of the system in the first test case scenario we use an image which contained 100 small letters as shown in Fig. 6.11. The training set is constructed through two images containing 40 examples of each small letter in the German alphabet which took around 19.86 s. The results are presented in Tables 6.2 and 6.3. The parameter C is regularization parameter, ρ is bias term, κ is kernel function, σ is smoothing function which reduces variance and ϕ is mapping function in feature space. Further discussion on these parameters is available in [6].

For the next test case scenario we use for training only the features corresponding to capital letters. The image used for testing contained 100 letters as shown in

aaqa äáää bbbb ßßßß cccc dddd eeee ffff

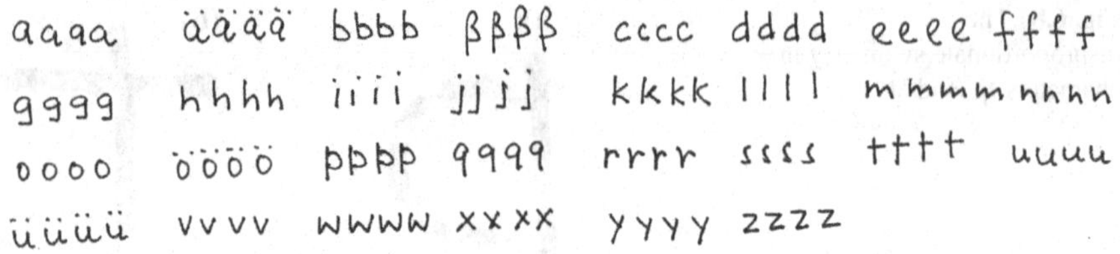

gggg hhhh iiii jjjj kkkk llll mmmm nhnn

oooo öööö pppp qqqq rrrr ssss tttt uuuu

üüüü vvvv wwww xxxx yyyy zzzz

Fig. 6.11 The test image for small letters

Table 6.2 The training set results corresponding to small letters (for FSVM)

Kernel Function	C	ρ	κ	σ	ϕ	Precision (%)
Linear	1	–	–	–	–	94
Linear	10	–	–	–	–	93
Linear	100	–	–	–	–	93
RBF	10	–	–	0.25	–	94
RBF	10	–	–	0.15	–	94
RBF	10	–	–	0.10	–	95
RBF	10	–	–	0.05	–	96
RBF	10	–	–	0.03	–	96
RBF	10	–	–	0.02	–	96
RBF	10	–	–	0.01	–	97
RBF	10	–	–	0.005	–	97
Polynomial	10	2	2	–	–	97
Polynomial	10	2	4	–	–	96
Polynomial	10	2	1	–	–	96
Polynomial	10	2	0.5	–	–	94
Polynomial	10	3	2	–	–	93
Polynomial	10	3	4	–	–	93
Polynomial	10	3	1	–	–	95
Polynomial	10	3	0.5	–	–	96
Polynomial	10	4	2	–	–	95
Polynomial	10	4	4	–	–	93
Polynomial	10	4	1	–	–	95
Polynomial	10	4	0.5	–	–	95
Sigmoid	10	–	0.5	–	1	94
Sigmoid	10	–	0.5	–	5	93
Sigmoid	10	–	0.2	–	1	94
Sigmoid	10	–	0.7	–	1	95

Table 6.3 The training set results corresponding to small letters (for FRSVM)

Kernel Function	C	ρ	κ	σ	ϕ	Precision (%)
Linear	1	–	–	–	–	95
Linear	10	–	–	–	–	94
Linear	100	–	–	–	–	94
RBF	10	–	–	0.25	–	95
RBF	10	–	–	0.15	–	95
RBF	10	–	–	0.10	–	96
RBF	10	–	–	0.05	–	96
RBF	10	–	–	0.03	–	98
RBF	10	–	–	0.02	–	97
RBF	10	–	–	0.01	–	98
RBF	10	–	–	0.005	–	98
Polynomial	10	2	2	–	–	98
Polynomial	10	2	4	–	–	98
Polynomial	10	2	1	–	–	98
Polynomial	10	2	0.5	–	–	96
Polynomial	10	3	2	–	–	94
Polynomial	10	3	4	–	–	93
Polynomial	10	3	1	–	–	95
Polynomial	10	3	0.5	–	–	95
Polynomial	10	4	2	–	–	95
Polynomial	10	4	4	–	–	95
Polynomial	10	4	1	–	–	95
Polynomial	10	4	0.5	–	–	95
Sigmoid	10	–	0.5	–	1	95
Sigmoid	10	–	0.5	–	5	95
Sigmoid	10	–	0.2	–	1	96
Sigmoid	10	–	0.7	–	1	96

Fig. 6.12. The training set is constructed through two images containing 40 examples of each capital letter in the German alphabet which took around 19.86 s. The results are presented in Tables 6.4 and 6.5.

For the final test case scenario we use for training the features corresponding to both small and capital letters. The images used for testing are the ones used in the first and second test cases. The training set construction took around 19.86 s. The results are presented in Tables 6.6 and 6.7.

A comparative performance of the soft computing techniques (RFMLP, FSVM, FRSVM, HFBRNN) used for German language with the traditional techniques (MLP, SVM) is provided in Fig. 6.13 for samples of 30 datasets. It is to be noted that the architecture of MLP used for classification is 70–100–100–52 [7] and sigmoid kernel is used with SVM.

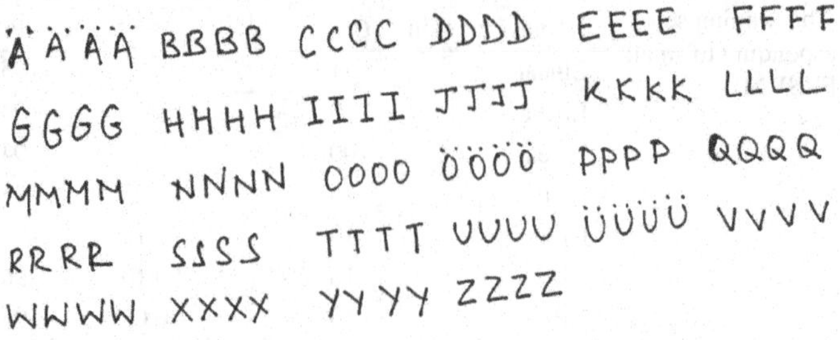

Fig. 6.12 The test image for capital letters

Table 6.4 The training set results corresponding to capital letters (for FSVM)

Kernel Function	C	ρ	κ	σ	ϕ	Precision (%)
Linear	1	–	–	–	–	95
Linear	10	–	–	–	–	96
Linear	100	–	–	–	–	94
RBF	10	–	–	0.25	–	97
RBF	10	–	–	0.15	–	95
RBF	10	–	–	0.10	–	97
RBF	10	–	–	0.05	–	94
RBF	10	–	–	0.03	–	93
RBF	10	–	–	0.02	–	93
RBF	10	–	–	0.01	–	94
RBF	10	–	–	0.005	–	94
Polynomial	10	2	2	–	–	89
Polynomial	10	2	4	–	–	94
Polynomial	10	2	1	–	–	93
Polynomial	10	2	0.5	–	–	94
Polynomial	10	3	2	–	–	94
Polynomial	10	3	4	–	–	89
Polynomial	10	3	1	–	–	96
Polynomial	10	3	0.5	–	–	93
Polynomial	10	4	2	–	–	96
Polynomial	10	4	4	–	–	94
Polynomial	10	4	1	–	–	95
Polynomial	10	4	0.5	–	–	93
Sigmoid	10	–	0.5	–	1	89
Sigmoid	10	–	0.2	–	1	89

Table 6.5 The training set results corresponding to capital letters (for FRSVM)

Kernel Function	C	ρ	κ	σ	ϕ	Precision (%)
Linear	1	–	–	–	–	95
Linear	10	–	–	–	–	95
Linear	100	–	–	–	–	93
RBF	10	–	–	0.25	–	97
RBF	10	–	–	0.15	–	96
RBF	10	–	–	0.10	–	98
RBF	10	–	–	0.05	–	95
RBF	10	–	–	0.03	–	94
RBF	10	–	–	0.02	–	93
RBF	10	–	–	0.01	–	93
RBF	10	–	–	0.005	–	95
Polynomial	10	2	2	–	–	93
Polynomial	10	2	4	–	–	95
Polynomial	10	2	1	–	–	94
Polynomial	10	2	0.5	–	–	95
Polynomial	10	3	2	–	–	96
Polynomial	10	3	4	–	–	95
Polynomial	10	3	1	–	–	97
Polynomial	10	3	0.5	–	–	93
Polynomial	10	4	2	–	–	96
Polynomial	10	4	4	–	–	96
Polynomial	10	4	1	–	–	93
Polynomial	10	4	0.5	–	–	95
Sigmoid	10	–	0.5	–	1	94
Sigmoid	10	–	0.2	–	1	95

Table 6.6 The training set results corresponding to both small and capital letters (for FSVM)

Kernel Function	C	ρ	κ	σ	ϕ	Precision (%)
Linear	1	–	–	–	–	89
Linear	10	–	–	–	–	89
RBF	10	–	–	0.25	–	93
RBF	10	–	–	0.10	–	93
RBF	10	–	–	0.05	–	89
RBF	10	–	–	0.01	–	89
Polynomial	10	2	2	–	–	88
Polynomial	10	3	2	–	–	89
Polynomial	10	4	2	–	–	94
Sigmoid	10	–	0.5	–	–	93
Sigmoid	10	–	0.2	–	–	93

Table 6.7 The training set results corresponding to both small and capital letters (for FRSVM)

Kernel Function	C	ρ	κ	σ	ϕ	Precision (%)
Linear	1	–	–	–	–	89
Linear	10	–	–	–	–	93
RBF	10	–	–	0.25	–	93
RBF	10	–	–	0.10	–	95
RBF	10	–	–	0.05	–	89
RBF	10	–	–	0.01	–	89
Polynomial	10	2	2	–	–	89
Polynomial	10	3	2	–	–	94
Polynomial	10	4	2	–	–	96
Sigmoid	10	–	0.5	–	–	94
Sigmoid	10	–	0.2	–	–	95

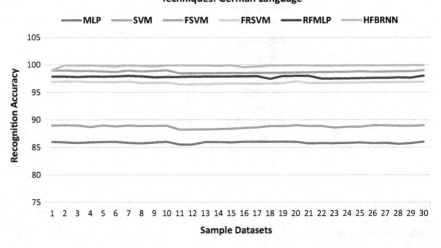

Fig. 6.13 The comparative performance of soft computing versus traditional techniques for German language

All the tests are conducted on PC having Intel P4 processor with 4.43 GHz, 512 MB DDR RAM @ 400 MHz with 512 kB cache. The training set construction was the longest operation of the system where the processor was loaded to 25% and the application occupied around 54.16 MB of memory. During idle mode the application consumes 43.02 MB of memory.

6.8.3 Hierarchical Fuzzy Bidirectional Recurrent Neural Networks

The experimental results for a subset of the German characters both uppercase and lowercase using HFBRNN are shown in the Table 6.8. The German characters can be categorized as ascenders (characters such as c, C etc.) and descenders (characters such as p, Ä etc.) as illustrated in Sect. 6.8.1. HFBRNN shows better results than the traditional methods as well RFMLP [7]. The testing algorithm is applied to all standard cases of characters [26] in various samples. An accuracy of around 99.9% has been achieved in all the cases.

Table 6.8 The experimental results for a subset of the German characters using HFBRNN

Characters	Successful recognition (%)	Unsuccessful recognition (%)	No recognition (%)
a	99.9	0.1	0
ä	99.9	0.1	0
b	99.9	0.1	0
β	99.9	0.1	0
c	99.9	0.1	0
d	99.9	0.1	0
e	99.9	0.1	0
f	99.9	0.1	0
g	99.9	0.1	0
h	99.9	0.1	0
i	99.9	0.1	0
j	99.9	0.1	0
k	99.9	0.1	0
l	99.9	0.1	0
m	99.9	0.1	0
n	99.9	0.1	0
o	99.9	0.1	0
ö	99.9	0.1	0
p	99.9	0.1	0
q	99.9	0.1	0
r	99.9	0.1	0
s	99.9	0.1	0
t	99.9	0.1	0
u	99.9	0.1	0
ü	99.9	0.1	0
u	99.9	0.1	0
w	99.9	0.1	0
x	99.9	0.1	0

(continued)

Table 6.8   (continued)

Characters	Successful recognition (%)	Unsuccessful recognition (%)	No recognition (%)
y	99.9	0.1	0
z	99.9	0.1	0
A	99.9	0.1	0
Ä	99.9	0.1	0
B	99.9	0.1	0
C	99.9	0.1	0
D	99.9	0.1	0
E	99.9	0.1	0
F	99.9	0.1	0
G	99.9	0.1	0
H	99.9	0.1	0
I	99.9	0.1	0
J	99.9	0.1	0
K	99.9	0.1	0
L	99.9	0.1	0
M	99.9	0.1	0
N	99.9	0.1	0
O	99.9	0.1	0
Ö	99.9	0.1	0
P	99.9	0.1	0
Q	99.9	0.1	0
R	99.9	0.1	0
S	99.9	0.1	0
T	99.9	0.1	0
U	99.9	0.1	0
Ü	99.9	0.1	0
V	99.9	0.1	0
w	99.9	0.1	0
X	99.9	0.1	0
Y	99.9	0.1	0
Z	99.9	0.1	0

6.9   Further Discussions

All the feature based classification based methods used in this chapter give better results than the traditional approaches [6]. On comparing with other algorithms, it is observed that RFMLP work with high accuracy in almost all cases which include intersections of loop and instances of multiple crossings. This algorithm focussed on the processing of various asymmetries in the German characters. RFMLP function consider the conditions: (a) the validity of the algorithms for non-cursive German alphabets. (b) the requirement of the algorithm that height of both upper and lower case characters to be proportionally same and (c) for extremely illegible handwriting the accuracy achieved by the algorithm is very less.

FSVM and FRSVM also give superior performance compared to traditional SVM [6]. FSVM and FRSVM achieve a precision rate up to 97% in case of training with sets corresponding to either small or capital letters and up to 98% in case of training with sets corresponding to both small and capital letters respectively. Thus the system achieved its goal through the recognition of characters from an image. The future research direction entails in expanding the system through addition of techniques which determine automatically the optimal parameters of kernel functions. Further, any version of SVM can be used to better perform the feature based classification task. We are also experimenting with other robust versions of SVM which will improve the overall recognition accuracy of the German characters.

HFBRNN produces the best results in terms of accuracy for all cases including loop intersections and multiple crossing instances. This algorithm also focussed on the processing of various asymmetries in the German characters as RFMLP. HFBRNN achieves high successful recognition rate of about 99.9% for all the German characters.

Finally we conclude the chapter with a note that we are exploring further results on InftyCDB-2 dataset using other soft computing techniques such as fuzzy markov random fields and rough version of hierarchical bidirectional recurrent neural networks.

References

1. Bunke, H., Wang, P. S. P. (Editors), Handbook of Character Recognition and Document Image Analysis, World Scientific, 1997.
2. Chaudhuri, A., Ghosh, S. K., Sentiment Analysis of Customer Reviews Using Robust Hierarchical Bidirectional Recurrent Neural Network, Book Chapter: Artificial Intelligence Perspectives in Intelligent Systems, Radek Silhavy, Roman Senkerik, Zuzana Kominkova Oplatkova, Petr Silhavy, Zdenka Prokopova, (Editors), Advances in Intelligent Systems and Computing, Springer International Publishing, Switzerland, Volume 464, pp 249 – 261, 2016.
3. Chaudhuri, A., Fuzzy Rough Support Vector Machine for Data Classification, International Journal of Fuzzy System Applications, 5(2), pp 26 – 53, 2016.
4. Chaudhuri, A., Modified Fuzzy Support Vector Machine for Credit Approval Classification, AI Communications, 27(2), pp 189 – 211, 2014.
5. Chaudhuri, A., De, Fuzzy Support Vector Machine for Bankruptcy Prediction, Applied Soft Computing, 11(2), pp 2472 – 2486, 2011.
6. Chaudhuri, A., Applications of Support Vector Machines in Engineering and Science, Technical Report, Birla Institute of Technology Mesra, Patna Campus, India, 2011.
7. Chaudhuri, A., Some Experiments on Optical Character Recognition Systems for different Languages using Soft Computing Techniques, Technical Report, Birla Institute of Technology Mesra, Patna Campus, India, 2010.

8. Chaudhuri, A., De, K., Job Scheduling using Rough Fuzzy Multi-Layer Perception Networks, Journal of Artificial Intelligence: Theory and Applications, 1(1), pp 4 – 19, 2010.

9. Chaudhuri, A., De, K., Chatterjee, D., Discovering Stock Price Prediction Rules of Bombay Stock Exchange using Rough Fuzzy Multi-Layer Perception Networks, Book Chapter: Forecasting Financial Markets in India, Rudra P. Pradhan, Indian Institute of Technology Kharagpur, (Editor), Allied Publishers, India, pp 69 – 96, 2009.

10. Cheriet, M., Kharma, N., Liu, C. L., Suen, C. Y., Character Recognition Systems: A Guide for Students and Practitioners, John Wiley and Sons, 2007.

11. De, R. K., Basak, J., Pal, S. K., Neuro-Fuzzy Feature Evaluation with Theoretical Analysis, Neural Networks, 12(10), pp 1429 – 1455, 1999.

12. De, R. K., Pal, N. R., Pal, S. K., Feature Analysis: Neural Network and Fuzzy Set Theoretic Approaches, Pattern Recognition, 30(10), pp 1579 – 1590, 1997.

13. Gonzalez, R. C., Woods, R. E., Digital Image Processing, 3rd Edition, Pearson, 2013.

14. Jain, A. K., Duin, R. P. W., Mao, J., Statistical Pattern Recognition: A Review, IEEE Transactions on Pattern Analysis and Machine Intelligence, 22(1), pp 4 – 37, 2000.

15. Jain, A. K., Fundamentals of Digital Image Processing, Prentice Hall, India, 2006.

16. Pal, S. K., Mitra, S., Mitra, P., Rough-Fuzzy Multilayer Perception: Modular Evolution, Rule Generation and Evaluation, IEEE Transactions on Knowledge and Data Engineering, 15(1), pp 14 – 25, 2003.

17. Russ, J. C., The Image Processing Handbook, CRC Press, 6th Edition, 2011.

18. Schantz, H. F., The History of OCR, Recognition Technology Users Association, Manchester Centre, VT, 1982.

19. Taghva, K., Borsack, J., Condit, A., Effects of OCR Errors on Ranking and Feedback using the Vector Space Model, Information Processing and Management, 32(3), pp 317 – 327, 1996.

20. Taghva, K., Borsack, J., Condit, A., Evaluation of Model Based Retrieval Effectiveness with OCR Text, ACM Transactions on Information Systems, 14(1), pp 64 – 93, 1996.

21. Taghva, K., Borsack, J., Condit, A., Erva, S., The Effects of Noisy Data on Text Retrieval, Journal of American Society for Information Science, 45 (1), pp 50 – 58, 1994.

22. Young, T. Y., Fu, K. S., Handbook of Pattern Recognition and Image Processing, Academic Press, 1986.

23. Zadeh, L. A., Fuzzy Sets, Information and Control, 8(3), pp 338 – 353, 1965.

24. Zimmermann, H. J., Fuzzy Set Theory and its Applications, 4th Edition, Kluwer Academic Publishers, Boston, 2001.

25. https://en.wikipedia.org/wiki/German_language.

26. http://www.inftyproject.org/en/database.html.

27. https://www.abbyy.com/finereader/.

28. https://www.csie.ntu.edu.tw/~cjlin/libsvm/.

Chapter 7
Optical Character Recognition Systems for Latin Language

Abstract The optical character recognition (OCR) systems for Latin language were the most primitive ones and occupy a significant place in pattern recognition. The Latin language OCR systems have been used successfully in a wide array of commercial applications. The different challenges involved in the OCR systems for Latin language is investigated in this Chapter. The pre-processing activities such as text region extraction, skew detection and correction, binarization, noise removal, character segmentation and thinning are performed on the datasets considered. The feature extraction is performed through fuzzy Genetic Algorithms (GA). The feature based classification is performed through important soft computing techniques viz rough fuzzy multilayer perceptron (RFMLP), fuzzy support vector machine (FSVM) and fuzzy rough support vector machine (FRSVM) and hierarchical fuzzy bidirectional recurrent neural networks (HFBRNN). The superiority of soft computing techniques is demonstrated through the experimental results.

Keywords Latin language OCR · RFMLP · FSVM · FRSVM · HFBRNN

7.1 Introduction

In pattern recognition the optical character recognition (OCR) for Latin language [1, 2] has carved a niche place for itself. This has led to the development of several OCR systems for the Latin language [3]. Latin is one of the ancient languages spoken in Europe particularly in Italy. The Latin language OCR systems have been used successfully in a wide array of commercial products [1]. Like the English, French and German languages the character recognition of Latin language has a high potential in data and word processing. Some commonly used applications of the OCR systems of Latin language [1] are automated postal address and ZIP code reading, data acquisition in bank checks, processing of archived institutional records etc.

© Springer International Publishing AG 2017   165
A. Chaudhuri et al., *Optical Character Recognition Systems for Different Languages with Soft Computing*, Studies in Fuzziness and Soft Computing 352,
DOI 10.1007/978-3-319-50252-6_7

The standardization of OCR character set for Latin language was provided through ISO/IEC 8859-1 [4]. ISO/IEC 8859-1 is a joint ISO and IEC series of standards for 8-bit coded character set for information interchange and development in cooperation with ASCII since 1987. The characters in the ISO/IEC 8859-1 basic character set are invariant characters. This was following the similar specifications as those used by the ISO basic German alphabets. It is the basis for most popular 8-bit character sets including Windows-1252 and the first block of characters in Unicode. The Windows-1252 codepage coincides with ISO-8859-1 for all codes except the range 128–159 (hex 80–9F) where the little-used C1 controls are replaced with additional characters including all the missing characters provided by ISO-8859-15. In the recent years, the OCR for Latin language has gained a considerable importance as the need for converting the scanned images into computer recognizable formats such as text documents has variety of applications. The Latin language based OCR systems is thus one of the most fascinating and challenging areas of pattern recognition with various practical applications [1].

The OCR process for any language involves extraction of defined characteristics called features to classify an unknown character into one of the known classes [1, 3, 5, 6] to a user defined accuracy level. As such any good OCR system is best defined in terms of the success of feature extraction and classification tasks. The same is true for the Latin language. The process becomes tedious in case the language has dependencies where some characters look identical. Thus the classification becomes a big challenge.

In this chapter we start the investigation of OCR systems considering the different challenges involved in Latin language. The different pre-processing activities such as text region extraction, skew detection and correction, binarization, noise removal, character segmentation and thinning are performed on the considered datasets [7]. The feature extraction is performed through fuzzy genetic algorithms. The feature based classification is performed through important soft computing techniques viz rough fuzzy multilayer perceptron (RFMLP) [8–10] two support vector machine (SVM) based methods such as fuzzy support vector machine (FSVM) [11, 12] and fuzzy rough support vector machine (FRSVM) [13] and hierarchical fuzzy bidirectional recurrent neural networks (HFBRNN) [14]. The experimental results demonstrate the superiority of soft computing techniques over the traditional methods.

This chapter is structured as follows. In Sect. 7.2 a brief discussion about the Latin language script and datasets used for experiments are presented. The different challenges of OCR for Latin language are highlighted in Sect. 7.3. The next section illustrates the data acquisition. In Sect. 7.5 different pre-processing activities on the datasets such as text region extraction, skew detection and correction, binarization, noise removal, character segmentation and thinning are presented. This is followed by a discussion of feature extraction on Latin language dataset in Sect. 7.6. The Sect. 7.7 explains the state of art of OCR for Latin language in terms of feature based classification methods. The corresponding experimental results are given in Sect. 7.8. Finally Sect. 7.9 concludes the chapter with some discussions and future research directions.

7.2 Latin Language Script and Experimental Dataset

In this section we present brief information about the Latin language script and the dataset used for experiments. Latin is one of the oldest languages of the world [4]. Latin was originally spoken in Latium, Italy. Through the power of the Roman Republic, it became the dominant language initially in Italy and subsequently throughout the Roman Empire. The vulgar version of Latin developed into the languages such as Italian, Portuguese, Spanish, French and Romanian. Many words to the English language have been contributed by Latin language. Latin and ancient Greek roots are used in theology, biology, and medicine. Latin is the official language of Vatican City Holy See along with Italian, Croatian and Polish Parliaments. It is a widely spoken language in different parts of South America and Europe.

By the late Roman Republic, the old version of Latin language has been standardized into classical Latin. The vulgar version of Latin language was the colloquial form spoken during the same time and attested in inscriptions and the works of comic playwrights like Plautus and Terence. The current Latin language is written from the 3rd century and Medieval Latin the language used from the 9th century to the Renaissance which used the Renaissance Latin. Later, the early modern Latin evolved. Latin has been used as the language of international communication, scholarship and science until the 18th century when it began to be supplanted by vernaculars. The ecclesiastical Latin remains the official language of the Holy See and the Roman Rite of the Catholic Church. Today many students, scholars and members of the Catholic clergy speak Latin fluently. Latin language is taught in primary, secondary and postsecondary educational institutions around the world. Latin is a highly inflected language with three distinct genders, seven noun cases, four verb conjugations, four verb principle parts, six tenses, three persons, three moods, two voices, two aspects and two numbers.

Latin is a classical language belonging to the Italic branch of the Indo-European languages [4]. The Latin alphabet is derived from the Etruscan and Greek alphabets. It is a superset of the German vocabulary. The Latin language is also notable for its wide spectrum of dialects with many unique varieties existing in the Americas and Europe.

The Latin language dataset used for performing OCR experiments is the ISO Latin-1 dataset and is adapted from [7]. ISO Latin-1 database contains Latin text which is used here to for training and testing. The database contains unconstrained handwritten text which are scanned at a resolution of 300 dpi and saved as PNG images with 256 gray levels. The Fig. 7.1 shows a sample snapshot from the database. InftyCDB-2 database contains around few thousand Latin characters where few hundred writers contributed samples of their handwriting.

Further details are available at [7].

nbsp 10/00 nbsp #160	° 11/00 deg #176	À 12/00 Agrave #192	Ð 13/00 ETH #208	à 14/00 agrave #224	ð 15/00 eth #240
¡ 10/01 iexcl #161	± 11/01 plusmn #177	Á 12/01 Aacute #193	Ñ 13/01 Ntilde #209	á 14/01 aacute #225	ñ 15/01 ntilde #241
¢ 10/02 cent #162	² 11/02 sup2 #178	Â 12/02 Acirc #194	Ò 13/02 Ograve #210	â 14/02 acirc #226	ò 15/02 ograve #242
£ 10/03 pound #163	³ 11/03 sup3 #179	Ã 12/03 Atilde #195	Ó 13/03 Oacute #211	ã 14/03 atilde #227	ó 15/03 oacute #243
¤ 10/04 curren #164	´ 11/04 acute #180	Ä 12/04 Auml #196	Ô 13/04 Ocirc #212	ä 14/04 auml #228	ô 15/04 ocirc #244
¥ 10/05 yen #165	µ 11/05 micro #181	Å 12/05 Aring #197	Õ 13/05 Otilde #213	å 14/05 aring #229	õ 15/05 otilde #245
¦ 10/06 brvbar #166	¶ 11/06 para #182	Æ 12/06 AElig #198	Ö 13/06 Ouml #214	æ 14/06 aelig #230	ö 15/06 ouml #246
§ 10/07 sect #167	· 11/07 middot #183	Ç 12/07 Ccedil #199	× 13/07 times #215	ç 14/07 ccedil #231	÷ 15/07 divide #247
¨ 10/08 uml #168	¸ 11/08 cedil #184	È 12/08 Egrave #200	Ø 13/08 Oslash #216	è 14/08 egrave #232	ø 15/08 oslash #248
© 10/09 copy #169	¹ 11/09 sup1 #185	É 12/09 Eacute #201	Ù 13/09 Ugrave #217	é 14/09 eacute #233	ù 15/09 ugrave #249
ª 10/10 ordf #170	º 11/10 ordm #186	Ê 12/10 Ecirc #202	Ú 13/10 Uacute #218	ê 14/10 ecirc #234	ú 15/10 uacute #250
« 10/11 laquo #171	» 11/11 raquo #187	Ë 12/11 Euml #203	Û 13/11 Ucirc #219	ë 14/11 euml #235	û 15/11 ucirc #251
¬ 10/12 not #172	¼ 11/12 frac14 #188	Ì 12/12 Igrave #204	Ü 13/12 Uuml #220	ì 14/12 igrave #236	ü 15/12 uuml #252
10/13 shy #173	½ 11/13 frac12 #189	Í 12/13 Iacute #205	Ý 13/13 Yacute #221	í 14/13 iacute #237	ý 15/13 yacute #253
® 10/14 reg #174	¾ 11/14 frac34 #190	Î 12/14 Icirc #206	Þ 13/14 THORN #222	î 14/14 icirc #238	þ 15/14 thorn #254
¯ 10/15 macr #175	¿ 11/15 iquest #191	Ï 12/15 Iuml #207	ß 13/15 szlig #223	ï 14/15 iuml #239	ÿ 15/15 yuml #255

Fig. 7.1 A sample text snapshot from ISO Latin-1 database

7.3 Challenges of Optical Character Recognition Systems for Latin Language

After English, French and German languages the OCR for Latin language has become one of the most successful applications of technology in pattern recognition and artificial intelligence. The OCR for Latin language has been the topic of active research since past few decades [1]. The most commercially available OCR system for Latin language is i2OCR [15] which supports major language formats and multi column document analysis. Considering the important aspects of versatility, robustness and efficiency, the commercial OCR systems are generally divided into four generations [1] as highlighted in Chap. 2. It is to be noted that this categorization refers also to the OCRs of Latin language.

Despite decades of research and existence of established commercial OCR products based on Latin language, the output from such OCR processes often contains errors. The more highly degraded is input, the greater is error rate. Since inputs form the first stage in a pipeline where later stages are designed to support sophisticated information extraction and exploitation applications, it is important to understand the effects of recognition errors on downstream analysis routines. Few questions are required to be addressed in this direction. They are as follows:

(a)   Are all recognition errors equal in impact or some are worse than others?
(b)   Can the performance of each stage be optimized in isolation or the end-to-end system should be considered?
(c)   In balancing the trade-off between the risk of over and under segmenting characters during OCR where should the line be drawn to maximize overall performance?

The answers to these questions often influence the way OCR systems for Latin language are designed and build for analysis [1].

The Latin language OCR system converts numerous published books in Latin language into editable computer text files. The latest research in this area has grown to incorporate some new methodologies to overcome the complexity of Latin writing style. All these algorithms have still not been tested for complete characters of Latin alphabet. Hence, there is a quest for developing an OCR system which handles all classes of Latin text and identify characters among these classes increasing versatility, robustness and efficiency in commercial OCR systems. The recognition of printed Latin characters is itself a challenging problem since there is a variation of the same character due to change of fonts or introduction of different types of noises. There may be noise pixels that are introduced due to scanning of the image. A significant amount of research has been done towards text data processing in Latin language from noisy sources [1]. The majority of the work has focused predominately on errors that arise during speech recognition systems [16, 17]. Several research papers have appeared which examines the noise problem from variety of perspectives with most emphasizing issues that are inherent in written and spoken Latin language [1]. However, there has been less work concentrating on noise induced by OCR. Some earlier works by [18] show that moderate error rates have little impact on effectiveness of traditional information retrieval measures. However, this conclusion is tied to certain assumptions about information retrieval through bag of words, OCR error rate which may not be too high and length of documents which may not be too short. Some other notable research works in this direction are given in [16, 19]. All these works try to address some significant issues involved in OCR systems for Latin language such as error correction, performance evaluation etc. involving flexible and rigorous mathematical treatment [1]. Besides this any Latin character can be represented in variety of fonts and sizes as per the needs and requirements of application. Further the character with same font and size may also be bold face character as well as normal one [9]. Thus the width of stroke also significantly affects recognition process. Therefore, a good character recognition approach for Latin language [1, 17, 20, 21]:

(a) Must eliminate noise after reading binary image data
(b) Smooth image for better recognition
(c) Extract features efficiently
(d) Train the system and
(e) Classify patterns accordingly.

7.4 Data Acquisition

The progress in automatic character recognition systems in Latin language is motivated from two categories according to the mode of data acquisition which can be either online or offline character recognition systems [1]. Following the lines of English language, the data acquisition of Latin language can be either online or offline character recognition systems. The offline character recognition captures data from paper through optical scanners or cameras whereas online recognition systems utilize digitizers which directly capture writing with the order of strokes, speed, pen-up and pen-down information. As such the scope of this text is restricted to OCR systems, we confine our discussion to offline character recognition [1] for Latin language. The Latin language datasets used in this research is mentioned in Sect. 7.2.

7.5 Data Pre-processing

Once the data has been acquired properly we proceed to pre-process the data. In pre-processing stage [21–23] a series of operations are performed here which includes text region extraction, skew detection and correction, binarization, noise removal, character segmentation and thinning or skeletonization. The main objective of pre-processing is to organize information so that the subsequent character recognition task becomes simpler. It essentially enhances the image rendering it suitable for segmentation.

7.5.1 Text Region Extraction

The text region extraction is used here as the first step in character recognition process. The input image $I_{P \times Q}$ is first partitioned into m number of blocks $B_i; i = 1, 2, \ldots, m$ such that:

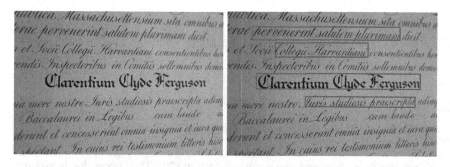

Fig. 7.2 The camera captured image and the text regions extracted from it

$$B_i \cap B_j = \emptyset \tag{7.1}$$

$$I_{P \times Q} = \bigcup_{i=1}^{m} B_i \tag{7.2}$$

A block B_i is a set of pixels represented as $B_i = [f(x, y)]_{H \times W}$ where H and W are the height and the width of the block respectively. Each individual block B_i is classified as either information block or background block based on the intensity variation within it. After removal of background blocks adjacent or contiguous information blocks constitute isolated components called as regions R_i; $i = 1, 2, \ldots, n$ such that:

$$R_i \cap R_j = \emptyset \,\forall \text{ but } \bigcup_{i=1}^{n} R_i \neq I_{P \times Q} \tag{7.3}$$

This is because some background blocks have been removed. The area of a region is always a multiple of the area of the blocks. These regions are then classified as text region or non-text region using various characteristics features of textual and non-textual regions such as dimensions, aspect ratio, information pixel density, region area, coverage ratio, histogram, etc. A detail description of this technique has been presented in [1]. The Fig. 7.2 shows a camera captured image and the text regions extracted from it.

7.5.2 Skew Detection and Correction

When a text document is fed into scanner either mechanically or manually a few degrees of tilt or skew is unavoidable. In skew angle the text lines in digital image make angle with horizontal direction. A number of methods are available in literature for identifying image skew angles [3]. They are basically categorized on the basis of projection profile analysis, nearest neighbor clustering, Hough transform, cross correlation and morphological transforms. The camera captured images very

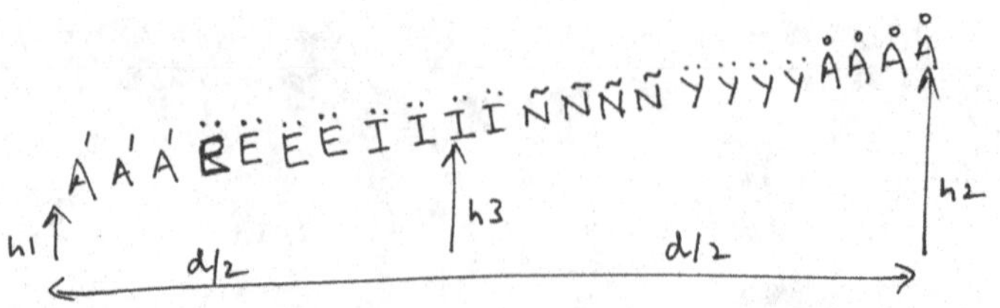

Fig. 7.3 The calculation of skew angle from *bottom* profile of a text region

often suffer from skew and perspective distortion [20]. They occur due to non-parallel axes or planes at the time of capturing the image. The acquired image does not become uniformly skewed mainly due to perspective distortion. The skewness of different portions of the image may vary between $+\alpha$ to $-\beta$ degrees where both α and β are positive numbers. Hence, the image cannot be deskewed at a single pass. On the other hand the effect of perspective distortion is distributed throughout the image. Its effect is hardly visible within a small region for example, the area of a character of the image. At the same time the image segmentation module generates only a few text regions. These text regions are deskewed using a computationally efficient and fast skew correction technique presented in [1].

Every text region basically has two types of pixels viz dark and gray. The dark pixels constitute the texts and the gray pixels are background around the texts. For the four sides of virtual bounding rectangle of a text region, there are four sets of values known as profiles. If the length and breadth of the bounding rectangle are M and N respectively, then two profiles will have M values each and the other two will have N values each. These values are the distances in terms of pixel from a side to the first gray or black pixel of the text region. Among these four profiles, the one which is from the bottom side of the text region is taken into consideration for estimating skew angle as shown in Fig. 7.3. This bottom profile is denoted as $\{h_i; i = 1, 2, \ldots, M\}$.

The mean $\mu = \frac{1}{M} \sum_{i=1}^{M} h_i$ and the first order moment $\tau = \frac{1}{M} \sum_{i=1}^{M} |\mu - h_i|$ values are calculated. Then, the profile size is reduced by excluding some h_i values that are not within the range $\mu \pm \tau$. The central idea behind this exclusion is that these elements hardly contribute to the actual skew of the text region. Now from the remaining profile, the elements viz leftmost h_1, rightmost h_2 and the middle one h_3 are chosen. The final skew angle is computed by averaging the three skew angles obtained from three pairs $h_1 - h_3$, $h_3 - h_2$ and $h_1 - h_2$. Once the skew angle for a text region is estimated it is rotated by the same angle.

7.5.3 Binarization

Binarization [21, 22] is the next step in character recognition process. A large number of binarization techniques are available in the literature [1] each of which is appropriate to particular image types. Its goal is to reduce the amount of information present in the image and keep only the relevant information. Generally the binarization techniques of gray scale images are classified into two categories viz overall threshold where single threshold is used in the entire image to form two classes (text and background) and local threshold where values of thresholds are determined locally (pixel-by-pixel or region-by-region). Here the skew corrected text region is binarized using an efficient binarization technique [17]. The algorithm is given below:

Binarization Algorithm:
begin
  for all pixels in (x, y) in *Text Region*
    if intensity $(x, y) < (G_{max} + G_{min})/2$
      then mark (x, y) as foreground
    else if number of foreground neighbors > 4
      then mark (x, y) as foreground
    else mark (x, y) as background
    end if
    end if
  end for
end

This is an improved version of bernsen's binarization method [1]. The arithmetic mean of maximum G_{max} and minimum G_{min} gray levels around a pixel is taken as the threshold for binarizing the pixel. In the present algorithm the eight immediate neighbors around the pixel subject to binarization are also taken as deciding factors for binarization. This approach is especially useful to connect the disconnected foreground pixels of a character.

7.5.4 Noise Removal

The scanned text documents often contain noise that arises due to printer, scanner, print quality, document age etc. Therefore, it is necessary to filter noise [22] before the image is processed. Here a low-pass filter is used to process the image [20] which is used for later processing. The main objective in the design of a noise filter is that it should remove as much noise as possible while retaining the entire signal [21].

7.5.5 Character Segmentation

Once the text image is skew corrected, binarized and noise removed, the actual text content is extracted. This process leads to character segmentation [22]. The commonly used segmentation algorithms in this direction are connected

Fig. 7.4 **a** The horizontal
histogram of text regions for
their segmentation (skewed
text region and its horizontal
histogram). **b** The horizontal
histogram of text regions for
their segmentation (skew
corrected text region and its
horizontal histogram)

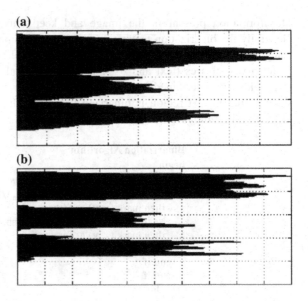

component labeling, x-y tree decomposition, run length smearing and Hough
transform [1]. After binarizing a noise free text region, the horizontal histogram
profile $\{f_i; i = 1, 2, \ldots, H_R\}$ of the region as shown in Fig. 7.4a, b is analyzed for
segmenting the region into text lines. Here f_i denotes the number of black pixel
along ith of the text region and hidden region denotes the height of the deskewed
text region. Firstly, all possible line segments are determined by thresholding the
profile values. The threshold is chosen so as to allow over segmentation. Text line
boundaries are referred by the values of i for which the value of f_i is less than
the threshold. Thus, n such segments represent $n - 1$ text lines. After that the inter
segment distances are analyzed and some segments are rejected based on the idea
that the distance between two lines in terms of pixels will not be too small and
the inter-segment distances are likely to become equal. A detail description of the
method is given in [20]. Using vertical histogram profile of each individual text
lines, words and characters are segmented. Sample segmented characters have
been shown in Fig. 7.5a, b.

7.5.6 Thinning

The character segmentation process is followed by thinning or skeletonization.
In thinning one-pixel-width representation or skeleton of an object is obtained by
preserving connectedness of the object and its end points [21]. The thinning process reduces image components to their essential information so that further analysis and recognition are facilitated. For instance, an alphabet can be handwritten
with different pens giving different stroke thicknesses but information presented

Fig. 7.5 **a** Skew correction and segmentation of text regions (an extracted text region). **b** Skew correction and segmentation of text regions (characters segmented from de-skewed text region)

(a)

(b)

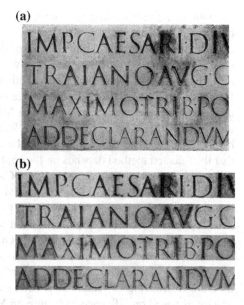

Fig. 7.6 An image before and after thinning

is same. This enables easier subsequent detection of pertinent features. As an illustration consider letter ё shown in Fig. 7.6 before and after thinning. A number of thinning algorithms have been used in the past with considerable success. The most common algorithm used is the classical hilditch algorithm and its variants [20]. Here hilditch algorithm is used for thinning [1]. For recognizing large graphical objects with filled regions which are often found in logos, region boundary detection is useful but for small regions corresponding to individual characters neither thinning nor boundary detection is performed. Rather entire pixel array representing the region is forwarded to subsequent stage of analysis.

7.6 Feature Selection Through Genetic Algorithms

As mentioned in Chap. 4 the heart of any OCR system is the formation of feature vector used in recognition stage. This fact is also valid for Latin language OCR system. This phase extracts the features from segmented areas of image containing characters to be recognized that distinguishes an area corresponding to a letter from an area corresponding to other letters. The feature extraction phase can thus be visualised as finding a set of parameters or features that define the shape of character as precise and unique. In Chap. 3 the term feature extraction is often used synonymously by feature selection which refers to algorithms that select the best subset of input feature set. These methods create new features based on transformations or combination of original features [1]. The features selected help in discriminating the characters. Achieving high recognition performance is attributed towards the selection of appropriate feature extraction methods. A large number of OCR based feature extraction methods are available in literature [1] except that the selected method depends on the application concerned. There is no universally accepted set of feature vectors in OCR. The features that capture topological and geometrical shape information are the most desired ones. The features that capture spatial distribution of black (text) pixels are also very important [22]. The genetic algorithm based feature extraction approach based on multilayer perceptron (MLP) is successfully applied for the OCR of Latin language [1].

A number of neural network and fuzzy set theoretic approaches [27, 28] have been proposed for feature analysis in recent past [1]. A feature quality index (FQI) measure for ranking of features has been suggested by [24]. The feature ranking process is based on influence of feature on MLP output. It is related to the importance of feature in discriminating among classes. The impact of qth feature on MLP output out of a total of p features is measured by setting feature value to zero for each input pattern. FQI is defined as the deviation of MLP output with qth feature value set to zero from output with all features present such that:

$$FQI_q = \frac{1}{n} \sum_{i=1}^{n} \left\| OV_i - OV_i^{(q)} \right\|^2 \tag{7.4}$$

In Eq. (7.4), OV_i and $OV_i^{(q)}$ are output vectors with all p features present and with qth feature set to zero. The features are ranked according to their importance as $q_1, \ldots, q_p$ if $FQI_{q_1} > \cdots > FQI_{q_p}$. In order to select best p' features from the set of p features, $\binom{p}{p'}$ possible subsets are tested one at a time. The quality index $FQI_k^{(p')}$ of kth subset S_k is measured as:

$$FQI_k^{(p')} = \frac{1}{n} \sum_{i=1}^{n} \left\| OV_i - OV_i^k \right\|^2 \tag{7.5}$$

In Eq. (7.5), OV_i^k is MLP output vector with x_i^k as input where x_i^k is derived from x_i as:

$$x_{ij}^k = \begin{cases} 0 & \text{if } j \in S_k \\ x_{ij} & \text{ow} \end{cases} \tag{7.6}$$

A subset S_j is selected as optimal set of features if $FQI_j^{(p')} \geq FQI_k^{(p')} \forall k, k \neq j$. An important observation here is that value of p' should be predetermined and $\binom{p}{p'}$

number of possible choices are verified to arrive at the best feature set. It is evident that no a priori knowledge is usually available to select the value p' and an exhaustive search is to be made for all values p' with $p' = 1, \ldots, p$. The number of possible trials is $(2^p - 1)$ which is prohibitively large for high values of p. To overcome drawbacks of above method, best feature set is selected by using of genetic algorithms [1]. Let us consider mask vector M where $M_i \in \{0, 1\}; i = 1, \ldots, p$ and each feature element $q_i, i = 1, \ldots, n$ is multiplied by corresponding mask vector element before reaching MLP input such that $I_i = q_i M_i$. MLP inputs are then written as:

$$I_i = \begin{cases} 0 & \text{if } M_i = 0 \\ q_i & \text{ow} \end{cases} \qquad (7.7)$$

Thus, a particular feature q_i reaches MLP if corresponding mask element is one. To find sensitivity of a particular feature q_j, mask bit M_j is set to zero. With respect to the above discussions when kth subset of feature set $\{q_1, \ldots, q_p\}$ is selected, all corresponding mask bits are set to zero and rest are set to one. When feature set multiplied by these mask bits reaches MLP, the effect of setting features of subset S_k to zero is obtained. Then value of FQI_k is calculated. It is to be noted that kth subset thus chosen may contain any number of feature elements and not pre-specified p' number of elements. Starting with an initial population of strings representing mask vectors, genetic algorithm is used with reproduction, crossover and mutation operators to determine best value of objective function. The objective function is FQI value of feature set S_k selected with mask bits set to zero for specific features and is given by:

$$FQI_k = \frac{1}{n} \sum_{i=1}^{n} \left\| OV_i - OV_i^k \right\|^2 \qquad (7.8)$$

In this process, both the problems of predetermining value of p' and searching through $\binom{p}{p'}$ possible combinations for each value of p'. In genetic algorithm implementation, the process is started with 20 features generated from fuzzy Hough transform such that the number of elements in mask vector is also 20. After running genetic algorithm for sufficiently large number of generations, mask string with best objective function value is determined. The feature elements corresponding to mask bits zero are chosen as selected set of features. The parameters for genetic algorithms are determined in terms of chromosome length, population size, mutation probability and crossover probability. MLP is next trained with only selected feature set for classification. The number of features varied when required. The genetic algorithm based feature selection method is shown in Fig. 7.7.

Fig. 7.7 The feature
selection process using
genetic algorithm

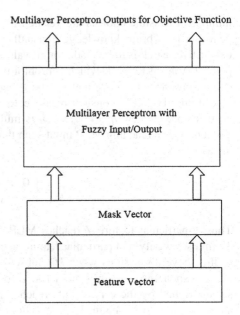

Multilayer Perceptron Outputs for Objective Function

Multilayer Perceptron with
Fuzzy Input/Output

Mask Vector

Feature Vector

7.7 Feature Based Classification: Sate of Art

After extracting the essential features from the pre-processed character image the concentration pointer turns on the feature based classification methods for OCR of Latin language. Considering the wide array of feature based classification methods for OCR, these methods are generally grouped into following four broad categories [19, 24, 25]:

(a) Statistical methods
(b) ANN based methods
(c) Kernel based methods
(d) Multiple classifier combination methods.

We discuss here the work done on Latin characters using some of the above-mentioned classification methods. All the classification methods used here are soft computing based techniques. The next three subsections highlight the feature based classification based on RFMLP [8–10], FSVM [11, 12], FRSVM [13] and HFBRNN [14] techniques. These methods are already discussed in Chap. 3. For further details interested readers can refer [1, 5].

7.7.1 Feature Based Classification Through Rough Fuzzy
         Multilayer Perceptron

The Latin language characters are recognized here through RFMLP [8–10] which is discussed in Chap. 3. For the Latin language RFMLP has been used with considerable success [1]. The diagonal feature extraction scheme is used for drawing of features from the characters. For classification stage we use these features extracted. RFMLP used here is a feed forward network with back propagation ANN having four hidden layers. The architecture used is 70–100–100–100–100– 100–52 [1] for classification. The four hidden layers and output layer uses tangent sigmoid as the activation function. The feature vector is again denoted by $F = (f_1, \ldots, f_m)$ where m denotes number of training images and each f has a length of 70 which represent the number of input nodes. The 52 neurons in the output layer correspond to the 26 modern Latin alphabets both uppercase and lowercase [7]. The network training parameters are briefly summarized as [1]:

(a) Input nodes: 70
(b) Hidden nodes: 100 at each layer
(c) Output nodes: 52
(d) Training algorithm: Scaled conjugate gradient backpropagation
(e) Performance function: Mean Square Error
(f) Training goal achieved: 0.000002
(g) Training epochs: 4000.

7.7.2 Feature Based Classification Through Fuzzy and Fuzzy Rough Support Vector Machines

In similar lines to the English language character recognition through FSVM and FRSVM in Sect. 4.7.3, FSVM [11, 12] and FRSVM [13] discussed in Chap. 3 is used here to recognize the Latin language characters. Over the past years SVM based methods have shown a considerable success in feature based classification [5]. With this motivation FSVM and FRSVM are used here for feature classification task. Both FSVM and FRSVM offer the possibility of selecting different types of kernel functions such as sigmoid, RBF, linear functions and determining the best possible values for these kernel parameters [5, 11, 12]. After selecting the kernel type and its parameters, FSVM and FRSVM are trained with the set of features obtained from other phases. Once the training gets over, FSVM and FRSVM are used to classify new character sets. The implementation is achieved through LibSVM library [26].

7.7.3 Feature Based Classification Through Hierarchical Fuzzy Rough Bidirectional Recurrent Neural Networks

After RFMLP, FSVM and FRSVM, the Latin language characters are recognized through HFBRNN [14] which is discussed in Chap. 3. For the Latin language HFBRNN has been used with considerable success [1]. HBRNN is a deep learning based technique [14]. In the recent past deep learning based techniques have shown a considerable success in feature based classification [14]. With this motivation HBRNN is used here for feature classification task and takes full advantage of deep recurrent neural network (RNN) towards modeling long-term information of data sequences. The recognition of characters done by HFBRNN at review level. The performance of HFBRNN is improved by fine tuning parameters of the network in a hierarchical fashion. The motivation is obtained from long short term memory (LSTM) and bidirectional LSTM (BLSTM) [14]. The evaluation is done on different types of highly biased character data. The implementation of HBRNN is performed in MATLAB [1].

7.8 Experimental Results

In this section, the experimental results for soft computing tools viz RFMLP, FSVM, FRSVM and HFBRNN on the Latin language dataset [7] highlighted in Sect. 7.2 are presented. The prima face is to select the best OCR system for Latin language [1].

7.8.1 Rough Fuzzy Multilayer Perceptron

The experimental results for a subset of the Latin characters both uppercase and lowercase using RFMLP are shown in the Table 7.1. Like the English characters, the Latin characters are categorized as ascenders (characters such as ç, Ç, ÿ etc.) and descenders (characters such as þ, Á, Ā etc.). RFMLP shows better results than the traditional methods [1]. The testing algorithm is applied to all standard cases of characters [7] in various samples. An accuracy of around 99% has been achieved in all the cases. However, after successful training and testing of algorithm the following flaws are encountered [1] which are identical to those encountered in English language:

(a) There may an instance when there is an occurrence of large and disproportionate symmetry in both ascenders as well as descenders as shown in Fig. 7.8a, b.
(b) There may be certain slants in certain characters which results in incorrect detection of characters as shown in Fig. 7.9.
(c) Sometimes one side of the image may have less white space and more pixel concentration whereas other side may have more white space and less pixel concentration. Due to this some characters are wrongly detected as shown in Fig. 7.10.

Table 7.1 The experimental results for a subset of the Latin characters using RFMLP Latin language

Characters	Successful recognition (%)	Unsuccessful recognition (%)	No recognition (%)
á	99	1	0
â	99	1	0
ã	99	1	0
ä	99	1	0
å	99	1	0
ç	99	1	0
è	98	1	1
é	99	1	0
ê	99	1	0
ë	99	1	0
ì	99	1	0
í	99	1	0
î	98	1	1
ï	98	1	1
ò	99	1	0
þ	99	1	0
û	99	1	0
ú	99	1	0
ù	99	1	0
ó	99	1	0
ô	99	1	0
õ	99	1	0
ö	98	1	1
ñ	98	1	1
ý	99	1	0
ÿ	99	1	0
Á	99	1	0
Â	99	1	0
Ã	99	1	0
Ä	99	1	0
Å	99	1	0
Ā	99%	1	0
Ç	99	1	0
È	99	1	0
É	99	1	0
Ê	99	1	0
Ë	99	1	0
Ì	99	1	0
Í	99	1	0
Î	98	1	1

(continued)

Table 7.1 (continued)

Characters	Successful recognition (%)	Unsuccessful recognition (%)	No recognition (%)
Ï	98	1	1
Ò	99	1	0
Ó	99	1	0
Ô	99	1	0
Õ	99	1	0
Ö	99	1	0
Ù	99	1	0
Ú	99	1	0
Ü	99	1	0
Ý	98	1	1
þ	98	1	1
Ð	99	1	0
B	99	1	0

(a)　　　　　　**(b)**

Fig. 7.8 The disproportionate symmetry in characters 'þ' and 'ï'

Fig. 7.9 The character 'ï' with slant

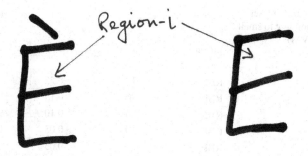

Fig. 7.10 The uneven pixel combination for character 'È' (region-i on the *left image* has less white space than the *right image*)

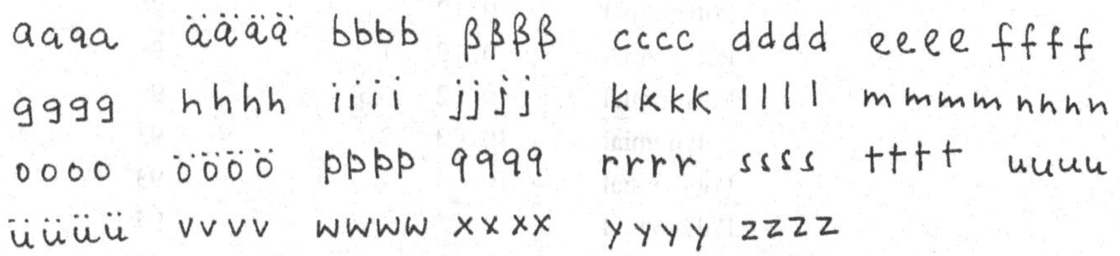

aaaa äáäà bbbb ßßßß cccc dddd eeee ffff

gggg hhhh iiii jjjj kkkk llll mmmm nnnn

oooo öòöò pppp qqqq rrrr ssss tttt uuuu

üüüü vvvv wwww xxxx yyyy zzzz

Fig. 7.11 The test image for small letters

7.8.2 *Fuzzy and Fuzzy Rough Support Vector Machines*

FSVM and FRSVM show promising results for feature classification task through different types of kernel functions by selecting the best possible values for kernel parameters [11–13]. For testing accuracy of the system in the first test case scenario we use an image which contained 100 small letters as shown in Fig. 7.11. The training set is constructed through two images containing 40 examples of each small letter in the Latin alphabet which took around 19.68 s. The results are presented in Tables 7.2 and 7.3. The parameter C is regularization parameter, ρ is bias term, κ is kernel function, σ is smoothing function which reduces variance and ϕ is mapping function in feature space. Further discussion on these parameters is available in [5].

For the next test case scenario we use for training only the features corresponding to capital letters. The image used for testing contained 100 letters as shown in Fig. 7.12. The training set is constructed through two images containing 40 examples of each capital letter in the Latin alphabet which took around 19.96 s. The results are presented in Tables 7.4 and 7.5.

For the final test case scenario we use for training the features corresponding to both small and capital letters. The images used for testing are the ones used in the first and second test cases. The training set construction took around 37.86 s. The results are presented in Tables 7.6 and 7.7.

Table 7.2 The training set results corresponding to small letters (for FSVM)

Kernel function	C	ρ	κ	σ	ϕ	Precision (%)
Linear	1	–	–	–	–	93
Linear	10	–	–	–	–	93
Linear	100	–	–	–	–	93
RBF	10	–	–	0.25	–	94
RBF	10	–	–	0.15	–	94
RBF	10	–	–	0.10	–	95
RBF	10	–	–	0.05	–	95
RBF	10	–	–	0.03	–	96
RBF	10	–	–	0.02	–	96
RBF	10	–	–	0.01	–	97
RBF	10	–	–	0.005	–	97
Polynomial	10	2	2	–	–	96
Polynomial	10	2	4	–	–	96
Polynomial	10	2	1	–	–	96
Polynomial	10	2	0.5	–	–	94
Polynomial	10	3	2	–	–	93
Polynomial	10	3	4	–	–	93
Polynomial	10	3	1	–	–	95
Polynomial	10	3	0.5	–	–	95
Polynomial	10	4	2	–	–	95
Polynomial	10	4	4	–	–	93
Polynomial	10	4	1	–	–	95
Polynomial	10	4	0.5	–	–	95
Sigmoid	10	–	0.5	–	1	93
Sigmoid	10	–	0.5	–	5	93
Sigmoid	10	–	0.2	–	1	94
Sigmoid	10	–	0.7	–	1	95

A comparative performance of the soft computing techniques (RFMLP, FSVM, FRSVM, HFBRNN) used for Latin language with the traditional techniques (MLP, SVM) is provided in Fig. 7.13 for samples of 30 datasets. It is to be noted that the architecture of MLP used for classification is 70–100–100–52 [1] and sigmoid kernel is used with SVM.

All the tests are conducted on PC having Intel P4 processor with 4.43 GHz, 512 MB DDR RAM @ 400 MHz with 512 KB cache. The training set construction was the longest operation of the system where the processor was loaded to 25% and the application occupied around 54.16 MB of memory. During idle mode the application consumes 43.02 MB of memory.

Table 7.3 The training set results corresponding to small letters (for FRSVM)

Kernel function	C	ρ	κ	σ	ϕ	Precision (%)
Linear	1	–	–	–	–	94
Linear	10	–	–	–	–	93
Linear	100	–	–	–	–	93
RBF	10	–	–	0.25	–	95
RBF	10	–	–	0.15	–	95
RBF	10	–	–	0.10	–	96
RBF	10	–	–	0.05	–	96
RBF	10	–	–	0.03	–	97
RBF	10	–	–	0.02	–	97
RBF	10	–	–	0.01	–	98
RBF	10	–	–	0.005	–	98
Polynomial	10	2	2	–	–	98
Polynomial	10	2	4	–	–	98
Polynomial	10	2	1	–	–	98
Polynomial	10	2	0.5	–	–	95
Polynomial	10	3	2	–	–	93
Polynomial	10	3	4	–	–	93
Polynomial	10	3	1	–	–	95
Polynomial	10	3	0.5	–	–	95
Polynomial	10	4	2	–	–	95
Polynomial	10	4	4	–	–	94
Polynomial	10	4	1	–	–	95
Polynomial	10	4	0.5	–	–	95
Sigmoid	10	–	0.5	–	1	95
Sigmoid	10	–	0.5	–	5	95
Sigmoid	10	–	0.2	–	1	96
Sigmoid	10	–	0.7	–	1	96

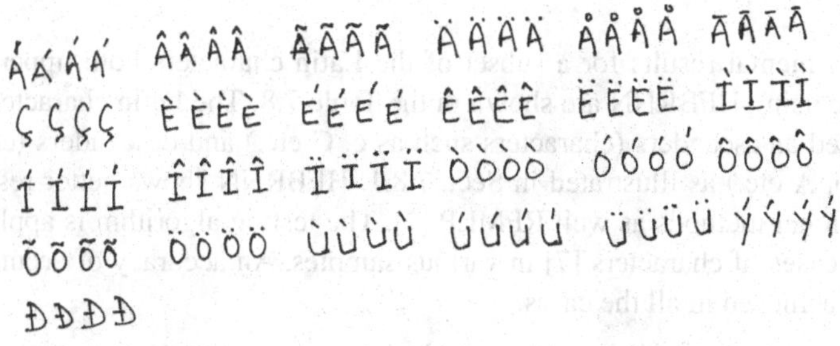

Fig. 7.12 The test image for capital letters

Table 7.4 The training set results corresponding to capital letters (for FSVM)

Kernel function	C	ρ	κ	σ	ϕ	Precision (%)
Linear	1	–	–	–	–	94
Linear	10	–	–	–	–	95
Linear	100	–	–	–	–	93
RBF	10	–	–	0.25	–	96
RBF	10	–	–	0.15	–	95
RBF	10	–	–	0.10	–	97
RBF	10	–	–	0.05	–	94
RBF	10	–	–	0.03	–	93
RBF	10	–	–	0.02	–	93
RBF	10	–	–	0.01	–	94
RBF	10	–	–	0.005	–	94
Polynomial	10	2	2	–	–	89
Polynomial	10	2	4	–	–	94
Polynomial	10	2	1	–	–	93
Polynomial	10	2	0.5	–	–	94
Polynomial	10	3	2	–	–	94
Polynomial	10	3	4	–	–	89
Polynomial	10	3	1	–	–	96
Polynomial	10	3	0.5	–	–	93
Polynomial	10	4	2	–	–	95
Polynomial	10	4	4	–	–	94
Polynomial	10	4	1	–	–	93
Polynomial	10	4	0.5	–	–	93
Sigmoid	10	–	0.5	–	1	89
Sigmoid	10	–	0.2	–	1	89

7.8.3 Hierarchical Fuzzy Rough Bidirectional Recurrent Neural Networks

The experimental results for a subset of the Latin characters both uppercase and lowercase using HFBRNN are shown in the Table 7.8. The Latin characters can be categorized as ascenders (characters such as c, C etc.) and descenders (characters such as p, Ä etc.) as illustrated in Sect. 7.8.1. HFBRNN shows better results than the traditional methods as well RFMLP [1]. The testing algorithm is applied to all standard cases of characters [7] in various samples. An accuracy of around 99.9% has been achieved in all the cases.

Table 7.5 The training set results corresponding to capital letters (for FRSVM)

Kernel function	C	ρ	κ	σ	φ	Precision (%)
Linear	1	–	–	–	–	95
Linear	10	–	–	–	–	95
Linear	100	–	–	–	–	93
RBF	10	–	–	0.25	–	97
RBF	10	–	–	0.15	–	96
RBF	10	–	–	0.10	–	98
RBF	10	–	–	0.05	–	95
RBF	10	–	–	0.03	–	94
RBF	10	–	–	0.02	–	93
RBF	10	–	–	0.01	–	93
RBF	10	–	–	0.005	–	95
Polynomial	10	2	2	–	–	93
Polynomial	10	2	4	–	–	95
Polynomial	10	2	1	–	–	94
Polynomial	10	2	0.5	–	–	95
Polynomial	10	3	2	–	–	96
Polynomial	10	3	4	–	–	94
Polynomial	10	3	1	–	–	97
Polynomial	10	3	0.5	–	–	93
Polynomial	10	4	2	–	–	96
Polynomial	10	4	4	–	–	95
Polynomial	10	4	1	–	–	93
Polynomial	10	4	0.5	–	–	93
Sigmoid	10	–	0.5	–	1	94
Sigmoid	10	–	0.2	–	1	93

Table 7.6 The training set results corresponding to both small and capital letters (for FSVM)

Kernel function	C	ρ	κ	σ	φ	Precision (%)
Linear	1	–	–	–	–	86
Linear	10	–	–	–	–	89
RBF	10	–	–	0.25	–	93
RBF	10	–	–	0.10	–	93
RBF	10	–	–	0.05	–	86
RBF	10	–	–	0.01	–	89
Polynomial	10	2	2	–	–	88
Polynomial	10	3	2	–	–	89
Polynomial	10	4	2	–	–	93
Sigmoid	10	–	0.5	–	–	93
Sigmoid	10	–	0.2	–	–	93

Table 7.7 The training set results corresponding to both small and capital letters (for FRSVM)

Kernel function	C	ρ	κ	σ	φ	Precision (%)
Linear	1	–	–	–	–	87
Linear	10	–	–	–	–	93
RBF	10	–	–	0.25	–	93
RBF	10	–	–	0.10	–	94
RBF	10	–	–	0.05	–	89
RBF	10	–	–	0.01	–	89
Polynomial	10	2	2	–	–	89
Polynomial	10	3	2	–	–	94
Polynomial	10	4	2	–	–	93
Sigmoid	10	–	0.5	–	–	94
Sigmoid	10	–	0.2	–	–	93

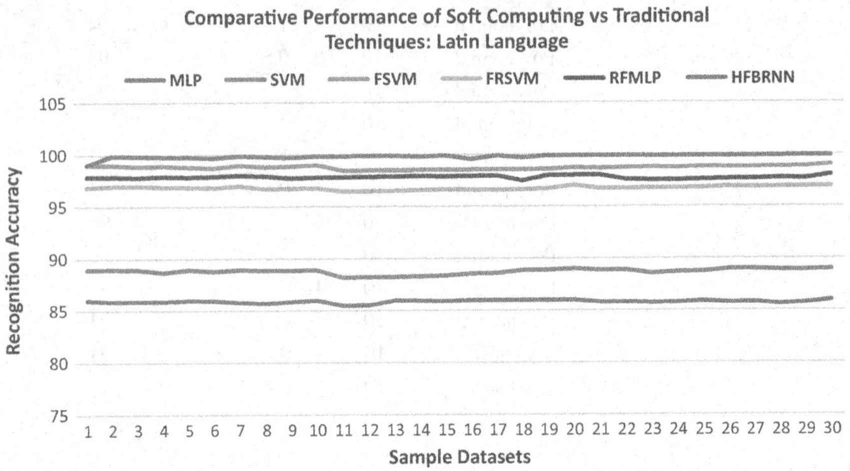

Fig. 7.13 The comparative performance of soft computing versus traditional techniques for Latin language

7.9 Further Discussions

All the feature based classification based methods used in this chapter give better results than the traditional approaches [5]. On comparing with other algorithms, it is observed that RFMLP work with high accuracy in almost all cases which include intersections of loop and instances of multiple crossings. This algorithm focussed on the processing of various asymmetries in the Latin characters. RFMLP function consider the conditions: (a) the validity of the algorithms for non-cursive Latin alphabets (b) the requirement of the algorithm that height of both upper and lower case characters to be proportionally same and (c) for extremely illegible handwriting the accuracy achieved by the algorithm is very less.

Table 7.8   The experimental results for a subset of the Latin characters using HFBRNN

Characters	Successful recognition (%)	Unsuccessful recognition (%)	No recognition (%)
á	99.9	0.1	0
â	99.9	0.1	0
ã	99.9	0.1	0
å	99.9	0.1	0
ä	99.9	0.1	0
ç	99.9	0.1	0
è	99.9	0.1	0
é	99.9	0.1	0
ê	99.9	0.1	0
ë	99.9	0.1	0
ì	99.9	0.1	0
í	99.9	0.1	0
î	99.9	0.1	0
ï	99.9	0.1	0
ò	99.9	0.1	0
þ	99.9	0.1	0
û	99.9	0.1	0
ú	99.9	0.1	0
ù	99.9	0.1	0
ó	99.9	0.1	0
ô	99.9	0.1	0
õ	99.9	0.1	0
ö	99.9	0.1	0
ñ	99.9	0.1	0
ý	99.9	0.1	0
ÿ	99.9	0.1	0
Á	99.9	0.1	0
Â	99.9	0.1	0
Ã	99.9	0.1	0
Ä	99.9	0.1	0
Å	99.9	0.1	0
A	99.9	0.1	0
Ç	99.9	0.1	0
È	99.9	0.1	0
É	99.9	0.1	0
Ê	99.9	0.1	0
Ë	99.9	0.1	0
Ì	99.9	0.1	0
Í	99.9	0.1	0
Î	99.9	0.1	0
Ï	99.9	0.1	0

(continued)

Table 7.8 (continued)

Characters	Successful recognition (%)	Unsuccessful recognition (%)	No recognition (%)
Ò	99.9	0.1	0
Ó	99.9	0.1	0
Ô	99.9	0.1	0
Õ	99.9	0.1	0
Ö	99.9	0.1	0
Ù	99.9	0.1	0
Ú	99.9	0.1	0
Ü	99.9	0.1	0
Ý	99.9	0.1	0
þ	99.9	0.1	0
Ð	99.9	0.1	0
B	99.9	0.1	0

FSVM and FRSVM also give superior performance compared to traditional SVM [5]. FSVM and FRSVM achieve a precision rate up to 97% in case of training with sets corresponding to either small or capital letters and up to 98% in case of training with sets corresponding to both small and capital letters respectively. Thus the system achieved its goal through the recognition of characters from an image. The future research direction entails in expanding the system through addition of techniques which determine automatically the optimal parameters of kernel functions. Further, any version of SVM can be used to better perform the feature based classification task. We are also experimenting with other robust versions of SVM which will improve the overall recognition accuracy of the Latin characters.

HFBRNN produces the best results in terms of accuracy for all cases including loop intersections and multiple crossing instances. This algorithm also focussed on the processing of various asymmetries in the Latin characters as RFMLP. HFBRNN achieves high successful recognition rate of about 99.9% for all the Latin characters.

Finally we conclude the Chapter with a note that we are exploring further results on InftyCDB-2 dataset using other soft computing techniques such as fuzzy markov random fields and rough version of hierarchical bidirectional recurrent neural networks.

References

1. Chaudhuri, A., Some Experiments on Optical Character Recognition Systems for different Languages using Soft Computing Techniques, Technical Report, Birla Institute of Technology Mesra, Patna Campus, India, 2010.
2. Schantz, H. F., The History of OCR, Recognition Technology Users Association, Manchester Centre, VT, 1982.

3. Bunke, H., Wang, P. S. P. (Editors), Handbook of Character Recognition and Document Image Analysis, World Scientific, 1997.
4. https://en.wikipedia.org/wiki/Latin_language.
5. Chaudhuri, A., Applications of Support Vector Machines in Engineering and Science, Technical Report, Birla Institute of Technology Mesra, Patna Campus, India, 2011.
6. Cheriet, M., Kharma, N., Liu, C. L., Suen, C. Y., Character Recognition Systems: A Guide for Students and Practitioners, John Wiley and Sons, 2007.
7. http://www.inftyproject.org/en/database.html.
8. Chaudhuri, A., De, K., Job Scheduling using Rough Fuzzy Multi-Layer Perception Networks, Journal of Artificial Intelligence: Theory and Applications, 1(1), pp 4–19, 2010.
9. Chaudhuri, A., De, K., Chatterjee, D., Discovering Stock Price Prediction Rules of Bombay Stock Exchange using Rough Fuzzy Multi-Layer Perception Networks, Book Chapter: Forecasting Financial Markets in India, Rudra P. Pradhan, Indian Institute of Technology Kharagpur, (Editor), Allied Publishers, India, pp 69–96, 2009.
10. Pal, S. K., Mitra, S., Mitra, P., Rough-Fuzzy Multilayer Perception: Modular Evolution, Rule Generation and Evaluation, IEEE Transactions on Knowledge and Data Engineering, 15(1), pp 14–25, 2003.
11. Chaudhuri, A., Modified Fuzzy Support Vector Machine for Credit Approval Classification, AI Communications, 27(2), pp 189–211, 2014.
12. Chaudhuri, A., De, Fuzzy Support Vector Machine for Bankruptcy Prediction, Applied Soft Computing, 11(2), pp 2472–2486, 2011.
13. Chaudhuri, A., Fuzzy Rough Support Vector Machine for Data Classification, International Journal of Fuzzy System Applications, 5(2), pp 26–53, 2016.
14. Chaudhuri, A., Ghosh, S. K., Sentiment Analysis of Customer Reviews Using Robust Hierarchical Bidirectional Recurrent Neural Network, Book Chapter: Artificial Intelligence Perspectives in Intelligent Systems, Radek Silhavy, Roman Senkerik, Zuzana Kominkova Oplatkova, Petr Silhavy, Zdenka Prokopova, (Editors), Advances in Intelligent Systems and Computing, Springer International Publishing, Switzerland, Volume 464, pp 249–261, 2016.
15. https://www.abbyy.com/finereader/.
16. Taghva, K., Borsack, J., Condit, A., Erva, S., The Effects of Noisy Data on Text Retrieval, Journal of American Society for Information Science, 45 (1), pp 50–58, 1994.
17. Young, T. Y., Fu, K. S., Handbook of Pattern Recognition and Image Processing, Academic Press, 1986.
18. Taghva, K., Borsack, J., Condit, A., Effects of OCR Errors on Ranking and Feedback using the Vector Space Model, Information Processing and Management, 32(3), pp 317–327, 1996.
19. Taghva, K., Borsack, J., Condit, A., Evaluation of Model Based Retrieval Effectiveness with OCR Text, ACM Transactions on Information Systems, 14(1), pp 64–93, 1996.
20. Jain, A. K., Fundamentals of Digital Image Processing, Prentice Hall, India, 2006.
21. Russ, J. C., The Image Processing Handbook, CRC Press, 6th Edition, 2011.
22. Gonzalez, R. C., Woods, R. E., Digital Image Processing, 3rd Edition, Pearson, 2013.
23. Jain, A. K., Duin, R. P. W., Mao, J., Statistical Pattern Recognition: A Review, IEEE Transactions on Pattern Analysis and Machine Intelligence, 22(1), pp 4–37, 2000.
24. De, R. K., Pal, N. R., Pal, S. K., Feature Analysis: Neural Network and Fuzzy Set Theoretic Approaches, Pattern Recognition, 30(10), pp 1579–1590, 1997.
25. De, R. K., Basak, J., Pal, S. K., Neuro-Fuzzy Feature Evaluation with Theoretical Analysis, Neural Networks, 12(10), pp 1429–1455, 1999.
26. https://www.csie.ntu.edu.tw/~cjlin/libsvm/.
27. Zadeh, L. A., Fuzzy Sets, Information and Control, 8(3), pp 338–353, 1965.
28. Zimmermann, H. J., Fuzzy Set Theory and its Applications, 4th Edition, Kluwer Academic Publishers, Boston, 2001.

Chapter 8
Optical Character Recognition Systems for Hindi Language

Abstract The optical character recognition (OCR) systems for Hindi language were the most primitive ones and occupy a significant place in pattern recognition. The Hindi language OCR systems have been used successfully in a wide array of commercial applications. The different challenges involved in the OCR systems for Hindi language is investigated in this Chapter. The pre-processing activities such as binarization, noise removal, skew detection, character segmentation and thinning performed on the datasets considered. The feature extraction is performed through fuzzy Hough transform. The feature based classification is performed through important soft computing techniques viz rough fuzzy multilayer perceptron (RFMLP), fuzzy support vector machine (FSVM), fuzzy rough support vector machine (FRSVM) and fuzzy markov random fields (FMRF). The superiority of soft computing techniques is demonstrated through the experimental results.

Keywords Hindi language OCR · RFMLP · FSVM · FRSVM · FMRF

8.1 Introduction

Several optical character recognition (OCR) systems have been developed for Hindi language [8]. Hindi is the most widely spoken language across the Indian subcontinent [30] and several other parts of the world after Mandarin, Spanish and English. This motivation has led to the development of OCR systems for Hindi language [3]. Hindi language OCR systems have been used successfully in a wide array of commercial applications [8]. The character recognition of Hindi language has a high potential in data and word processing among other Indian languages. Some commonly used applications of the OCR systems of Hindi language [8] are automated postal address and ZIP code reading, data acquisition in bank checks, processing of archived institutional records etc. as evident in other languages such as English, French, German and Hindi.

© Springer International Publishing AG 2017 193
A. Chaudhuri et al., *Optical Character Recognition Systems for Different Languages with Soft Computing*, Studies in Fuzziness and Soft Computing 352,
DOI 10.1007/978-3-319-50252-6_8

The first complete OCR in Hindi language was introduced by [19] where structural and template features are used for recognizing basic, modified and compound characters. To recognize real life printed character documents of varying size and font [2] have proposed statistical features. Later [16, 18] have used the gradient features for recognizing handwritten Hindi characters. [1] has used density, moment of curve and descriptive component for recognizing Hindi handwritten characters and numerals. [21] has proposed a set of primitives, such as global and local horizontal and vertical line segments, right and left slant and loop for recognizing handwritten Hindi characters. [22] has used directional chain code information of the contour points of the characters for recognizing handwritten Hindi characters.

The standardization of OCR character set for Hindi language was provided through ISCII (IS 13194:1991; earlier IS 13194:1988) [30] as shown in the Fig. 8.1. It is the national standard for Devanagari character set encoding based on earlier standard IS 10402:1982 [8]. ISCII is a standard for Devanagari script and has been used for other Indian languages. The standard contains ASCII in lower 128 slots and Devanagari alphabet superset in upper 128 slots and therefore it is a single byte standard. Though it is primarily an encoding standard and sorting is usually not catered directly in such standards, the standard was devised to do some implicit sorting directly on encoding. The variations of ISCII include PC-ISCII

	Hex	0	1	2	3	4	5	6	7	8	9	A	B	C	D	E	F
Hex	Dec	0	16	32	48	64	80	96	112	128	144	160	176	192	208	224	240
0	0	NUL	DLE	SP	0	@	P	`	p				ओ	ड	ऱ	ঁ	EXT
1	1	SOH	DC1	!	1	A	Q	a	q			ँ	ओ	ण	ल	ঁ	ॐ
2	2	STX	DC2	"	2	B	R	b	r			ं	आँ	त	ळ	ঁ	१
3	3	ETX	DC3	#	3	C	S	c	s			ः	क	थ	क़	ঁ	२
4	4	EOT	DC4	$	4	D	T	d	t			अ	ख	द	य	ओ	३
5	5	ENQ	NAK	%	5	E	U	e	u			आ	ग	ध	श	ओ	४
6	6	ACK	SYN	&	6	F	V	f	v			इ	घ	न	ष	ओ	५
7	7	BEL	ETB	'	7	G	W	g	w			ई	ङ	ऩ	स	ओ	६
8	8	BS	CAN	(	8	H	X	h	x			उ	च	प	ह	ॉ	७
9	9	HT	EM	)	9	I	Y	i	y			ऊ	छ	फ	INV	ॉ	८
A	10	LF	SUB	*	:	J	Z	j	z			ऋ	ज	ब	ऒ	।	९
B	11	VT	ESC	+	;	K	[	k	{			ऐ	झ	भ	ि		
C	12	FF	FS	,	<	L	\	l	|			ए	ञ	म	ो		
D	13	CR	GS	-	=	M	]	m	}			ऐ	ट	य	ॉ		
E	14	SO	RS	.	>	N	^	n	~			ऍ	ठ	र	ॉ		
F	15	SI	US	/	?	O	_	o	DEL			ओ	ड	र	ॉ	ATR	

Fig. 8.1 The ISCII code chart IS 13194:1991

and language specific ISCII charts [34]. The official standard publication is available in [36]. The unicode provides an international standard for Devanagari character set encoding based on IS 13194:1988 from 0900 till 097F and therefore is not exactly equivalent to IS 13194:1991 [33, 35]. This has been used for Hindi and other Devanagari script based languages including Marathi, Sanskrit, Prakrit, Sindhi, etc. There are other encodings which have been used by vendors which not prevalent anymore. There are also encoding converters available which can convert among various encodings and platforms [37, 38]. In the recent past, OCR for Hindi language has gained a sizeable momentum, as the need for converting the scanned images into computer recognizable formats such as text documents has grown considerably. The Hindi language based OCR systems is thus one of the most fascinating and challenging areas of Indian pattern recognition research community [8].

The OCR process for any language involves extraction of defined characteristics called features to classify an unknown character into one of the known classes [8, 11] to a user defined accuracy level. As such any good OCR system is best defined in terms of the success of feature extraction and classification tasks. The same is true for Hindi language. The process becomes tedious in case the language has dependencies where some characters look identical. Thus the classification becomes a big challenge.

In this chapter we start the investigation of OCR systems considering the different challenges involved in Hindi language. The different pre-processing activities such as binarization, noise removal, skew detection and correction, character segmentation and thinning are performed on the considered datasets [31]. The feature extraction is performed through discrete Hough transformation. The feature based classification is performed through important soft computing techniques viz rough fuzzy multilayer perceptron (RFMLP) [8–10, 17] two support vector machine (SVM) based methods such as fuzzy support vector machine (FSVM) [5, 6] and fuzzy rough support vector machine (FRSVM) [4, 7] and fuzzy markov random fields (FMRF) [28]. The experimental results demonstrate the superiority of soft computing techniques over the traditional methods.

This chapter is structured as follows. In Sect. 8.2 a brief discussion about the Hindi language script and datasets used for experiments are presented. The different challenges of OCR for Hindi language are highlighted in Sect. 8.3. The next section illustrates the data acquisition. In Sect. 8.5 different pre-processing activities on the datasets such as binarization, noise removal, skew detection and correction, character segmentation and thinning are presented. This is followed by a discussion of feature extraction on Hindi language dataset in Sect. 8.6. The Sect. 8.7 explains the state of art of OCR for Hindi language in terms of feature based classification methods. The corresponding experimental results are given in Sect. 8.8. Finally Sect. 8.9 concludes the chapter with some discussions and future research directions.

8.2 Hindi Language Script and Experimental Dataset

In this section we present brief information about the Hindi language script and the dataset used for experiments. Hindi language spelled as हिन्दी [30] is called the modern standard Hindi (मानक हिन्दी) [30]. It is the standardized and Sanskritized register of the Hindustani language. The Hindi language is written in the Devnagri script (देवनागरी लिपि) which consists of 11 vowels and 33 consonants as shown in Fig. 8.2 [30]. It is written from left to right and is an abugida as well.

Hindi is considered to be a direct descendant of Sanskrit through Prakrit and Apabhramsha. The Hindi language has been influenced by Dravidian, Turkish, Persian, Arabic, Portuguese and English. Some of the notable dialects of Hindi are Braj, Awadhi, Khariboli etc. The dialect of Hindustani on which Standard Hindi is based is Khariboli which is the vernacular of Delhi and the surrounding western Uttar Pradesh and southern Uttarakhand.

Hindi is the official language of the Union of India and the *lingua franca* of the Hindi belt languages [8, 30]. Hindi language is the member of the Indo Aryan group within the Indo Iranian branch of the Indo European language family. It is the preferred official language of India along with English and the other languages recognized in the Indian constitution. In India, Hindi is spoken as a first language by nearly 425 million people and as a second language by some 120 million more. Significant Hindi speaking communities are also found in South Africa, Mauritius, Bangladesh, Yemen and Uganda.

After independence in 1947, the government of India instituted some notable conventions for the Hindi language. These included the standardization of Hindi grammar and standardization of the Hindi orthography using the Devanagari script by the Central Hindi Directorate of the Ministry of Education and Culture. This effort was towards bringing uniformity in writing and improving the shape of some Devanagari characters. The Constituent Assembly adopted Hindi as an official language of India on 14 September 1949 which is celebrated as Hindi Day. At the state level of Union of India, Hindi is the official language of the some Indian

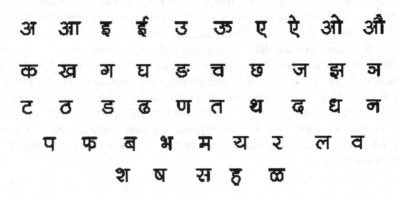

Fig. 8.2 The Hindi language consisting of 11 vowels and 33 consonants

states like Bihar, Chhattisgarh, Haryana, Himachal Pradesh, Jharkhand, Madhya Pradesh, Rajasthan, Uttar Pradesh and Uttarakhand. Hindi is also accorded the status of official language in the Union Territories like Andaman and Nicobar Islands, Chandigarh, Dadra and Nagar Haveli, Daman and Diu and National Capital Territory.

Outside Indian Union, Hindi is the official language in Fiji. The Hindi spoken in Fiji is Fiji Hindi, a form of Awadhi [30]. Hindi is also spoken by a large population of Madhesis of Nepal who are having roots in north India but have migrated to Nepal over hundreds of years ago. Being extremely close to Urdu, Hindi is easily understood by people having knowledge of Urdu. Apart from this, Hindi is spoken by the large Indian diaspora which hails from the Hindi speaking belt of India. A substantially large Indian diaspora lives in countries like United States, United Kingdom, Canada, United Arab Emirates, Saudi Arabia, Australia, South Africa and many other countries where it is natively spoken among the communities.

The Hindi language dataset used for performing OCR experiments is the HP Labs India Indic Handwriting dataset and is adapted from [31]. The database contains Hindi (Devanagari) text which is used here to for training and testing. The database contains unconstrained handwritten text which are scanned at a resolution of 300 dpi and saved as PNG images with 256 gray levels. The Fig. 8.3 shows a sample snapshot from the database. Further details are available at [31].

8.3   Challenges of Optical Character Recognition Systems for Hindi Language

The OCR for Hindi language has become one of the most successful applications in pattern recognition and artificial intelligence among the Indian languages after Bengali language [3, 8, 11]. It is rigorously perused after investigation topic by the Indian OCR research community [8]. The most commercially available OCR system for Hindi language is i2OCR [32]. Considering the important aspects of versatility, robustness and efficiency, the commercial OCR systems for Hindi language are generally divided into three generations [8]. The methods mentioned in Sect. 8.1 do not consider shape variation for extracting features. But in most of the Indian languages especially Hindi, a large number of similar shape type characters such as basic and conjunct are present. From this point of view, novel features based on the topography of a character has to be used to improve the performance of existing OCR in Hindi script. The major features worth mentioning are as follows:

(a) The main challenge to design an OCR for Hindi language is to handle large scale shape variation among different characters. The strokes in characters can be decomposed into segments which are straight lines, convexities or closed boundaries or hole. The topography of character strokes from 4 viewing directions and different convex shapes formed by the character strokes with respect to the presence of closed region boundaries need to be considered.

Fig. 8.3 The HP labs India
Indic handwriting Hindi
(Devanagari) dataset

(b) The extracted features are represented by a shape based graph where each node contains the topographic feature and they all are placed with respect to their centroids and relative positions in the original character image.
(c) This topographic feature set helps to differentiate very similar shape and type characters in a proper way.

Despite some rigorous research and existence of established commercial OCR products based on Hindi language, the output from such OCR processes often contains errors. The more highly degraded is input, the greater is error rate. Since inputs form the first stage in a pipeline where later stages are designed to support sophisticated information extraction and exploitation applications, it is important to understand the effects of recognition errors on downstream analysis routines. Few questions are required to be addressed in this direction. They are as follows:

(a) Are all recognition errors equal in impact or some are worse than others?
(b) Can the performance of each stage be optimized in isolation or the end-to-end system should be considered?
(c) In balancing the trade-off between the risk of over and under segmenting characters during OCR where should the line be drawn to maximize overall performance?

The answers to these questions often influence the way OCR systems for Hindi language are designed and build for analysis [11].

The Hindi language OCR system converts numerous published books in Hindi language into editable computer text files. The latest research in this area has grown to incorporate some new methodologies to overcome the complexity of Hindi writing style. All these algorithms have still not been tested for complete characters of Hindi alphabet. Hence, there is a quest for developing an OCR system which handles all classes of Hindi text and identify characters among these classes increasing versatility, robustness and efficiency in commercial OCR systems. The recognition of printed Hindi characters is itself a challenging problem since there is a variation of the same character due to change of fonts or introduction of different types of noises. There may be noise pixels that are introduced due to scanning of the image. A significant amount of research has been done towards text data processing in Hindi language from noisy sources [3]. The majority of the work has focused predominately on errors that arise during speech recognition systems [11]. Several research papers have appeared which examines the noise problem from variety of perspectives with most emphasizing issues that are inherent in written and spoken Hindi language [8]. However, there has been less work concentrating on noise induced by OCR. Some earlier works by [23] show that moderate error rates have little impact on effectiveness of traditional information retrieval measures. However, this conclusion is tied to certain assumptions about information retrieval through bag of words, OCR error rate which may not be too high and length of documents which may not be too short. Some other notable research works in this direction are given in [24, 25]. All these works try to address some significant issues involved in OCR systems for Hindi language such as error correction, performance evaluation etc. involving flexible and rigorous mathematical treatment [8]. Besides this any Hindi character can be represented in variety of fonts and sizes as per the needs and requirements of application. Further the character with same font and size may also be bold face character as well as normal one [11]. Thus the width of stroke also significantly affects recognition process. Therefore, a good character recognition approach for Hindi language [8, 15, 20, 26]:

(a)   Must eliminate noise after reading binary image data
(b)   Smooth image for better recognition
(c)   Extract features efficiently
(d)   Train the system and
(e)   Classify patterns accordingly.

8.4 Data Acquisition

The progress in automatic character recognition systems in Hindi language is generally evolved in two categories according to the mode of data acquisition which can be either online or offline character recognition systems. Offline character recognition captures data from paper through optical scanners or cameras whereas online recognition systems utilize digitizers which directly capture writing with the order of strokes, speed, pen-up and pen-down information. As such the scope of this text is restricted to OCR systems, we confine our discussion to offline character recognition [8] for Hindi language. The Hindi language datasets used in this research is mentioned in Sect. 8.2.

8.5 Data Pre-processing

Once the data has been acquired properly we proceed to pre-process the data. In pre-processing stage [8] a series of operations are performed viz binarization, noise removal, skew detection, character segmentation and thinning or skeletonization. The main objective of pre-processing is to organize information so that the subsequent character recognition task becomes simpler. It essentially enhances the image rendering it suitable for segmentation.

8.5.1 Binarization

Binarization [11] is an important first step in character recognition process. A large number of binarization techniques are available in the literature [3] each of which is appropriate to particular image types. Its goal is to reduce the amount of information present in the image and keep only the relevant information. Generally the binarization techniques of gray scale images are classified into two categories viz overall threshold where single threshold is used in the entire image to form two classes (text and background) and local threshold where values of thresholds are determined locally (pixel-by-pixel or region-by-region). Here, we calculate threshold Th of each pixel locally by [8]:

$$Th = (1 - k) \cdot m + k \cdot m + k \cdot \frac{\sigma}{R(m - M)} \tag{8.1}$$

In Eq. (8.1), k is set to 0.5; m is the average of all pixels in window; M is the minimum image grey level; σ is the standard deviation of all pixels in window and R is the maximum deviation of grayscale on all windows.

8.5.2 Noise Removal

The scanned text documents often contain noise that arises due to printer, scanner, print quality, document age etc. Therefore, it is necessary to filter noise [3] before the image is processed. Here a low-pass filter is used to process the image [8] which is used for later processing. The main objective in the design of a noise filter is that it should remove as much noise as possible while retaining the entire signal [11].

8.5.3 Skew Detection and Correction

When a text document is fed into scanner either mechanically or manually a few degrees of tilt or skew is unavoidable. In skew angle the text lines in digital image make angle with horizontal direction. A number of methods are available in literature for identifying image skew angles [3]. They are basically categorized on the basis of projection profile analysis, nearest neighbor clustering, Hough transform, cross correlation and morphological transforms. The aforementioned methods correct the detected skew. Some of the notable research works in this direction are available in [8]. Here Hough transform is used for skew detection and correction [8].

8.5.4 Character Segmentation

Once the text image is binarized, noise removed and skew corrected, the actual text content is extracted. This process leads to character segmentation [15, 20, 26]. The commonly used segmentation algorithms in this direction are connected component labeling, x-y tree decomposition, run length smearing and Hough transform [3]. Here Hough transform is used for character segmentation [8].

Fig. 8.4 An image before
and after thinning

8.5.5 *Thinning*

The character segmentation process is followed by thinning or skeletonization. In thinning one-pixel-width representation or skeleton of an object is obtained by preserving connectedness of the object and its end points [8]. The thinning process reduces image components to their essential information so that further analysis and recognition are facilitated. For instance, an alphabet can be handwritten with different pens giving different stroke thicknesses but information presented is same. This enables easier subsequent detection of pertinent features. As an illustration consider letter 'प' shown in Fig. 8.4 before and after thinning. A number of thinning algorithms have been used in the past with considerable success. The most common algorithm used is the classical hilditch algorithm and its variants [8]. Here hilditch algorithm is used for thinning [11]. For recognizing large graphical objects with filled regions which are often found in logos, region boundary detection is useful but for small regions corresponding to individual characters neither thinning nor boundary detection is performed. Rather entire pixel array representing the region is forwarded to subsequent stage of analysis.

8.6 Feature Extraction Through Hough Transform

As mentioned in Chap. 4 the heart of any OCR system is the formation of feature vector used in recognition stage. This fact is also valid for Hindi language OCR system. This phase extracts the features from segmented areas of image containing characters to be recognized that distinguishes an area corresponding to a letter from an area corresponding to other letters. The feature extraction phase can thus be visualised as finding a set of parameters or features that define the shape of character as precise and unique. In Chap. 3 the term feature extraction is often used synonymously by feature selection which refers to algorithms that select the best subset of input feature set. These methods create new features based on transformations or combination of original features [8]. The features selected help in discriminating the characters. Achieving high recognition performance is attributed towards the selection of appropriate feature extraction methods. A large number of OCR based feature extraction methods are available in literature [8] except that the selected method depends on the application concerned. There is no universally accepted set of feature vectors in OCR. The features that capture topological and geometrical shape information are the most desired ones. The features that capture spatial distribution of black (text) pixels are also very important [14]. The Hough transform based feature extraction approach is successfully applied for the OCR of Hindi language [8].

The Hough transform method is used for detection of lines and curves from images [15]. The basic Hough transform is generalized through fuzzy probabilistic concepts [8]. The fuzzy Hough transform treats image points as fuzzy points. The Hough transform for line detection uses mapping $r = x\cos\theta + y\sin\theta$ which provides three important characteristics of line in an image pattern. The parameters r and θ specify position and orientation of line. The count of (r, θ) accumulator cell used in Hough transform implementation specifies number of black pixels lying on it. With this motivation, a number of fuzzy sets on (r, θ) accumulator cells are defined. Some important fuzzy set definitions used here are presented in Table 8.1 for θ values in first quadrant. The definitions are extended for other values of θ. The fuzzy sets viz. long_line and short_line extract length information of different lines in pattern. The nearly_horizontal, nearly_vertical and slant_line represent

Table 8.1 Fuzzy set membership functions defined on Hough transform accumulator cells for line detection (x and y denote height and width of each character pattern)

Fuzzy Set	Membership Function
long_line	$\dfrac{cellcount}{\sqrt{x^2+y^2}}$
short_line	2(long_line) if $count \leq \sqrt{x^2+y^2}/2$ 2(1 − long_line) if $count > \sqrt{x^2+y^2}/2$
nearly_horizontal_line	$\dfrac{\theta}{90}$
nearly_vertical_line	1 − nearly_horizontal_line
slant_line	2(nearly_horizontal_line) if $\theta \leq 45$ 2(1 − nearly_horizontal_line) if $\theta > 45$
near_top	r/x if nearly_horizontal_line > nearly_vertical_line 0 otherwise
near_bottom	(1 − near_top) if nearly_horizontal_line > nearly_vertical_line 0 otherwise
near_vertical_centre	2(near_top) if ((nearly_horizontal_line > nearly_vertical_line) and $(r \leq x/2)$) 2(1 − near_top) if ((nearly_horizontal_line > nearly_vertical_line) and $(r > x/2)$) 0 otherwise
near_right_border	r/y if nearly_vertical_line > nearly_horizontal_line 0 otherwise
near_left_border	(1 − near_right_border) if nearly_vertical_line > nearly_horizontal_line 0 otherwise
near_horizontal_cen-tre	2(near_right_border) if ((nearly_vertical_line > nearly_horizontal_line) and $(r \leq y/2)$) 2(1 − near_right_border) if ((nearly_vertical_line > nearly_horizontal_line) and $(r > y/2)$) 0 otherwise

skew and near_top, near_bottom, near_vertical_centre, near_right, near_left and near_horizontal_centre extract position information of these lines. The characteristics of different lines in an image pattern are mapped into properties of these fuzzy sets. For interested readers further details are available in [8].

Based on basic fuzzy sets [27, 29], fuzzy sets are further synthesized to represent each line in pattern as combination of its length, position and orientation using t-norms [29]. The synthesized fuzzy sets are defined as long_slant_line $\equiv$ t-norm (slant_line, long_line), short_slant_line $\equiv$ t-norm (slant_line, short_line), nearly_vertical_long_line_near_left $\equiv$ t-norm (nearly_vertical_line, long_line, near_left_border). Similar basic fuzzy sets such as large_circle, dense_circle, centre_near_top etc. and synthesized fuzzy sets such as small_dense_circle_near_top, large_dense_circle_near_centre etc. are defined on (p, q, t) accumulator cells for circle extraction using Hough transform $t = \sqrt{(x - p)^2 + (y - q)^2}$. For a circle extraction (p, q) denotes origin, c is radius and count specifies the number of pixels lying on circle. A number of t-norms are available as fuzzy intersections among which standard intersection t-norm $(p, q) \equiv \min (p, q)$. For other pattern recognition problems suitable fuzzy sets may be similarly synthesized from basic sets of fuzzy Hough transform. A non-null support of synthesized fuzzy set implies presence of corresponding feature in a pattern. The height of each synthesized fuzzy set is chosen to define feature element and set of n such feature elements constitute an n-dimensional feature vector for a character.

8.7 Feature Based Classification: Sate of Art

After extracting the essential features from the pre-processed character image the concentration pointer turns on the feature based classification methods for OCR of Hindi language. Considering the wide array of feature based classification methods for OCR, these methods are generally grouped into following four broad categories [12, 13, 24]:

(a) Statistical methods
(b) ANN based methods
(c) Kernel based methods
(d) Multiple classifier combination methods.

We discuss here the work done on Hindi characters using some of the abovementioned classification methods. All the classification methods used here are soft computing based techniques. The next three subsections highlight the feature based classification based on RFMLP [8–10, 17], FSVM [5, 6], FRSVM [4, 7] and FMRF [28] techniques. These methods are already discussed in Chap. 3. For further details interested readers can refer [8].

8.7.1 Feature Based Classification Through Rough Fuzzy Multilayer Perceptron

The Hindi language characters are recognized here through RFMLP [8–10, 17] which is discussed in Chap. 3. For the Hindi language RFMLP has been used with considerable success [8]. The diagonal feature extraction scheme is used for drawing of features from the characters. For classification stage we use these features extracted. RFMLP used here is a feed forward network with back propagation ANN having four hidden layers. The architecture used is 70–100–100–100–100–100–52 [8] for classification. The four hidden layers and output layer uses tangent sigmoid as the activation function. The feature vector is again denoted by $F = (f_1, \ldots, f_m)$ where m denotes number of training images and each f has a length of 70 which represent the number of input nodes. The 44 neurons in the output layer correspond to the 11 vowels and 33 consonants of Hindi alphabets [30]. The network training parameters are briefly summarized as [8]:

(a) Input nodes: 70
(b) Hidden nodes: 100 at each layer
(c) Output nodes: 44
(d) Training algorithm: Scaled conjugate gradient backpropagation
(e) Performance function: Mean Square Error
(f) Training goal achieved: 0.000002
(g) Training epochs: 4000.

8.7.2 Feature Based Classification Through Fuzzy and Fuzzy Rough Support Vector Machines

In similar lines to the English language character recognition through FSVM and FRSVM in Sect. 4.7.3, FSVM [5, 6] and FRSVM [4, 7] discussed in Chap. 3 is used here to recognize the Hindi language characters. Over the past years SVM based methods have shown a considerable success in feature based classification [8]. With this motivation FSVM and FRSVM are used here for feature classification task. Both FSVM and FRSVM offer the possibility of selecting different types of kernel functions such as sigmoid, RBF, linear functions and determining the best possible values for these kernel parameters [5, 6]. After selecting the kernel type and its parameters, FSVM and FRSVM are trained with the set of features obtained from other phases. Once the training gets over, FSVM and FRSVM are used to classify new character sets. The implementation is achieved through LibSVM library [8].

8.7.3 Feature Based Classification Through Fuzzy Markov Random Fields

After RFMLP, FSVM and FRSVM, the Hindi language characters are recognized through FMRFs [28] which is discussed in Chap. 3. For the Hindi language FMRFs has been used with considerable success [8]. FMRFs are born by integrating type-2 (T2) fuzzy sets with markov random fields (MRFs) resulting in T2 FMRFs. The T2 FMRFs handles both fuzziness and randomness in the structural pattern representation. The T2 membership function has a 3–D structure in which the primary membership function describes randomness and the secondary membership function evaluates the fuzziness of the primary membership function. The MRFs represent statistical patterns structurally in terms of neighborhood system and clique potentials [28]. The T2 FMRFs defines the same neighborhood system as the classical MRFs. In order to describe the uncertain structural information in patterns, the fuzzy likelihood clique potentials are derived from T2 fuzzy Gaussian mixture models. The fuzzy prior clique potentials are penalties for the mismatched structures based on prior knowledge. The T2 FMRFs models the hierarchical character structures present in the language characters. For interested readers further details are available in [8]. The evaluation is done on different types of highly biased character data. The implementation of FMRFs is performed in MATLAB [8].

8.8 Experimental Results

In this section, the experimental results for soft computing tools viz RFMLP, FSVM, FRSVM and FMRF on the Hindi language dataset [31] highlighted in Sect. 8.2 are presented. The prima face is to select the best OCR system for the Hindi language [8].

8.8.1 Rough Fuzzy Multilayer Perceptron

The experimental results for a subset of the Hindi characters both vowels and consonants using RFMLP are shown in Table 8.2. Like the English characters, the Hindi characters are categorized as ascenders (characters such as अ, आ, उ, ऊ etc.) and descenders (characters such as ए, ऐ, इ, ई, प etc.). RFMLP shows better results than the traditional methods [8]. The testing algorithm is applied to all standard cases of characters [31] in various samples. An accuracy of around 99% has been achieved in all the cases. However, after successful training and testing of algorithm the following flaws are encountered [8]:

(a) There may an instance when there is an occurrence of large and disproportionate symmetry in both ascenders as well as descenders as shown in Fig. 8.5a, b.

Table 8.2 The experimental results for a subset of the Hindi characters using RFMLP

Characters	Successful Recognition (%)	Unsuccessful Recognition (%)	No Recognition (%)
अ	99	1	0
आ	99	1	0
इ	99	1	0
ई	99	1	0
उ	99	1	0
ऊ	99	1	0
ए	98	1	1
ऐ	99	1	0
ओ	99	1	0
औ	99	1	0
क	99	1	0
ख	99	1	0
ग	98	1	1
घ	98	1	1
ङ	99	1	0
च	99	1	0
छ	99	1	0
ज	99	1	0
झ	99	1	0
ञ	99	1	0
ट	99	1	0
ठ	99	1	0
ड	98	1	1
ढ	98	1	1
ण	99	1	0
थ	99	1	0
द	99	1	0
ध	99	1	0
न	99	1	0
प	99	1	0
फ	99	1	0
ब	99	1	0
भ	99	1	0
म	99	1	0
य	99	1	0
र	99	1	0
ल	99	1	0
ळ	99	1	0
व	99	1	0
ह	98	1	1

(continued)

Table 8.2 (continued)

Characters	Successful Recognition (%)	Unsuccessful Recognition (%)	No Recognition (%)
श	98	1	1
ष	99	1	0
स	99	1	0

(b)   There may be certain slants in certain characters which results in incorrect detection of characters as shown in Fig. 8.6.

(c)   Sometimes one side of the image may have less white space and more pixel concentration whereas other side may have more white space and less pixel concentration. Due to this some characters are wrongly detected as shown in Fig. 8.7.

8.8.2 Fuzzy and Fuzzy Rough Support Vector Machines

FSVM and FRSVM show promising results for feature classification task through different types of kernel functions by selecting the best possible values for kernel parameters [5–7]. For testing accuracy of the system in the first test case scenario we use an image which contained 100 letters as shown in Fig. 8.8. The training set is constructed through two images containing 40 examples of each vowel letter in the Hindi alphabet which took around 19.89 s. The results are presented in Tables 8.3 and 8.4. The parameter C is regularization parameter, ρ is bias term, κ is kernel function, σ is smoothing function which reduces variance and ϕ is mapping function in feature space. Further discussion on these parameters is available in [8].

A comparative performance of the soft computing techniques (RFMLP, FSVM, FRSVM, FMRF) used for Hindi language with the traditional techniques (MLP, SVM) is provided in Fig. 8.9 for samples of 30 datasets. It is to be noted that the architecture of MLP used for classification is 70–100–100–52 [7] and sigmoid kernel is used with SVM.

All the tests are conducted on PC having Intel P4 processor with 4.43 GHz, 512 MB DDR RAM @ 400 MHz with 512 kB cache. The training set construction was the longest operation of the system where the processor was loaded to 25% and the application occupied around 54.16 MB of memory. During idle mode the application consumes 43.02 MB of memory.

8.8.3 Fuzzy Markov Random Fields

The experimental results for a subset of the Hindi characters both vowels and consonants using FMRFs are shown in the Table 8.7. The Hindi characters are

Fig. 8.5 The disproportionate symmetry in characters 'ए' and 'इ'

(a)   **(b)**

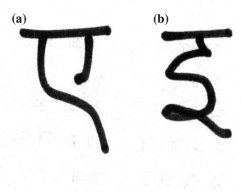

Fig. 8.6 The character 'ए' with slant

Fig. 8.7 The uneven pixel combination for character 'क' (region-i on the *left image* has less white space than the *right image*)

categorized as ascenders (characters such as अ, आ, उ, ऊ etc.) and descenders (characters such as ए, ऐ, इ, ई, प etc.) as illustrated in Sect. 8.8.1. FMRFs shows better results than the traditional methods shows better results than the traditional methods as well RFMLP [8]. The testing algorithm is applied to all standard cases of characters [31] in various samples. An accuracy of around 99.8% has been achieved in all the cases.

For the final test case scenario we use for training the features corresponding to Hindi letters. The images used for testing are the ones used in the first and second test cases. The training set construction took around 37.98 s. The results are presented in Tables 8.5 and 8.6.

8.9 Further Discussions

All the feature based classification based methods used in this Chapter give better results than the traditional approaches [7]. On comparing with other algorithms, it is observed that RFMLP work with high accuracy in almost all cases

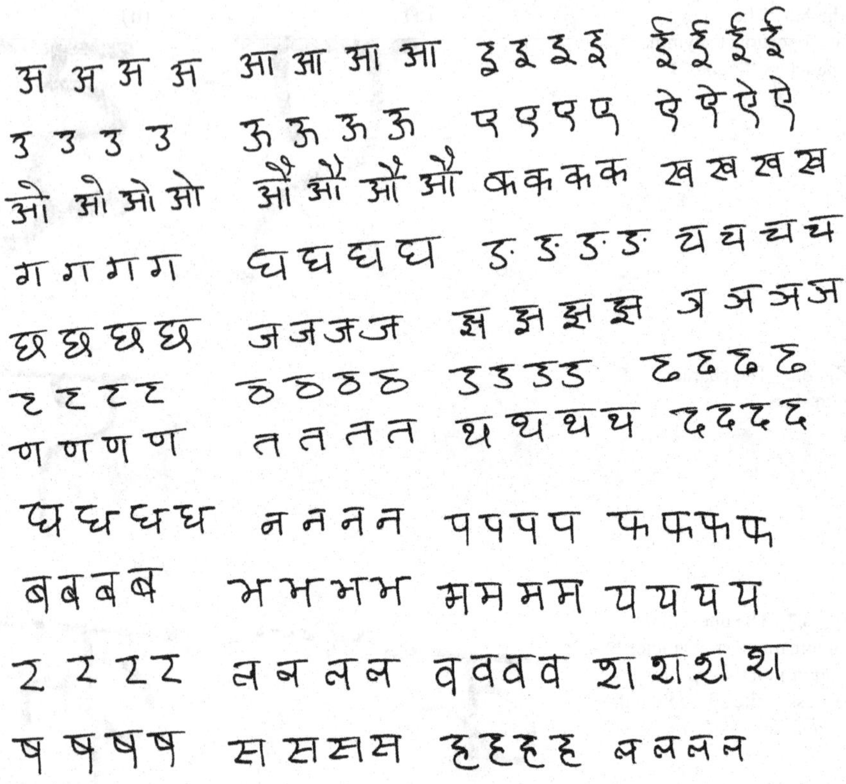

Fig. 8.8 The test image for Hindi letters

which include intersections of loop and instances of multiple crossings. This algorithm focussed on the processing of various asymmetries in the Hindi characters. RFMLP function consider the conditions: (a) the validity of the algorithms for non-cursive Hindi alphabets (b) the requirement of the algorithm that height of both upper and lower case characters to be proportionally same and (c) for extremely illegible handwriting the accuracy achieved by the algorithm is very less.

FSVM and FRSVM also give superior performance compared to traditional SVM [7]. FSVM and FRSVM achieve a precision rate up to 97% in case of training with sets corresponding to either small or capital letters and up to 98% in case of training with sets corresponding to both small and capital letters respectively. Thus the system achieved its goal through the recognition of characters from an image. The future research direction entails in expanding the system through addition of techniques which determine automatically the optimal parameters of kernel

Table 8.3 The training set results corresponding to Hindi letters (for FSVM)

Kernel Function	C	ρ	κ	σ	ϕ	Precision (%)
Linear	1	–	–	–	–	94
Linear	10	–	–	–	–	96
Linear	100	–	–	–	–	95
RBF	10	–	–	0.25	–	94
RBF	10	–	–	0.15	–	94
RBF	10	–	–	0.10	–	96
RBF	10	–	–	0.05	–	95
RBF	10	–	–	0.03	–	96
RBF	10	–	–	0.02	–	96
RBF	10	–	–	0.01	–	98
RBF	10	–	–	0.005	–	97
Polynomial	10	2	2	–	–	96
Polynomial	10	2	4	–	–	96
Polynomial	10	2	1	–	–	96
Polynomial	10	2	0.5	–	–	95
Polynomial	10	3	2	–	–	93
Polynomial	10	3	4	–	–	93
Polynomial	10	3	1	–	–	95
Polynomial	10	3	0.5	–	–	95
Polynomial	10	4	2	–	–	95
Polynomial	10	4	4	–	–	94
Polynomial	10	4	1	–	–	96
Polynomial	10	4	0.5	–	–	95
Sigmoid	10	–	0.5	–	1	93
Sigmoid	10	–	0.5	–	5	93
Sigmoid	10	–	0.2	–	1	94
Sigmoid	10	–	0.7	–	1	96

functions. Further, any version of SVM can be used to better perform the feature based classification task. We are also experimenting with other robust versions of SVM which will improve the overall recognition accuracy of the Hindi characters.

FMRF produces the best results in terms of accuracy for all cases including loop intersections and multiple crossing instances. This algorithm also focussed on the processing of various asymmetries in the Hindi characters as RFMLP. FMRF achieves high successful recognition rate of about 99.8% for all the Hindi characters.

Significant work is being done on Hindi language processing in Indian language technology centers at universities, Centre for Development of Advanced Computing (CDAC) and Technology Development for Indian Language (TDIL)

Table 8.4 The training set results corresponding to Hindi letters (for FRSVM)

Kernel Function	C	ρ	κ	σ	ϕ	Precision (%)
Linear	1	–	–	–	–	94
Linear	10	–	–	–	–	96
Linear	100	–	–	–	–	93
RBF	10	–	–	0.25	–	95
RBF	10	–	–	0.15	–	95
RBF	10	–	–	0.10	–	96
RBF	10	–	–	0.05	–	96
RBF	10	–	–	0.03	–	97
RBF	10	–	–	0.02	–	97
RBF	10	–	–	0.01	–	98
RBF	10	–	–	0.005	–	98
Polynomial	10	2	2	–	–	98
Polynomial	10	2	4	–	–	98
Polynomial	10	2	1	–	–	98
Polynomial	10	2	0.5	–	–	95
Polynomial	10	3	2	–	–	95
Polynomial	10	3	4	–	–	93
Polynomial	10	3	1	–	–	95
Polynomial	10	3	0.5	–	–	95
Polynomial	10	4	2	–	–	95
Polynomial	10	4	4	–	–	94
Polynomial	10	4	1	–	–	95
Polynomial	10	4	0.5	–	–	95
Sigmoid	10	–	0.5	–	1	95
Sigmoid	10	–	0.5	–	5	96
Sigmoid	10	–	0.2	–	1	96
Sigmoid	10	–	0.7	–	1	97

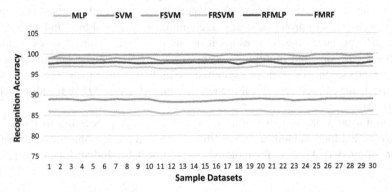

Fig. 8.9 The comparative performance of soft computing versus traditional techniques for Hindi language

Table 8.5 The training set results corresponding to Hindi letters (for FSVM)

Kernel Function	C	ρ	κ	σ	ϕ	Precision (%)
Linear	1	–	–	–	–	86
Linear	10	–	–	–	–	89
RBF	10	–	–	0.25	–	93
RBF	10	–	–	0.10	–	93
RBF	10	–	–	0.05	–	86
RBF	10	–	–	0.01	–	89
Polynomial	10	2	2	–	–	88
Polynomial	10	3	2	–	–	89
Polynomial	10	4	2	–	–	94
Sigmoid	10	–	0.5	–	–	93
Sigmoid	10	–	0.2	–	–	96

Table 8.6 The training set results corresponding to Hindi letters (for FRSVM)

Kernel Function	C	ρ	κ	σ	ϕ	Precision (%)
Linear	1	–	–	–	–	87
Linear	10	–	–	–	–	93
RBF	10	–	–	0.25	–	93
RBF	10	–	–	0.10	–	96
RBF	10	–	–	0.05	–	89
RBF	10	–	–	0.01	–	89
Polynomial	10	2	2	–	–	89
Polynomial	10	3	2	–	–	94
Polynomial	10	4	2	–	–	93
Sigmoid	10	–	0.5	–	–	94
Sigmoid	10	–	0.2	–	–	96

Table 8.7 The experimental results for a subset of the Hindi characters using FMRFs

Characters	Successful Recognition (%)	Unsuccessful Recognition (%)	No Recognition (%)
अ	99.8	0.2	0
आ	99.8	0.2	0
इ	99.8	0.2	0
ई	99.8	0.2	0
उ	99.8	0.2	0
ऊ	99.8	0.2	0
ए	99.8	0.2	0
ऐ	99.8	0.2	0
ओ	99.8	0.2	0
औ	99.8	0.2	0
क	99.8	0.2	0

(continued)

Table 8.7 (continued)

Characters	Successful Recognition (%)	Unsuccessful Recognition (%)	No Recognition (%)
ख	99.8	0.2	0
ग	99.8	0.2	0
घ	99.8	0.2	0
ङ	99.8	0.2	0
च	99.8	0.2	0
छ	99.8	0.2	0
ज	99.8	0.2	0
झ	99.8	0.2	0
ञ	99.8	0.2	0
ट	99.8	0.2	0
ठ	99.8	0.2	0
ड	99.8	0.2	0
ढ	99.8	0.2	0
ण	99.8	0.2	0
थ	99.8	0.2	0
द	99.8	0.2	0
ध	99.8	0.2	0
न	99.8	0.2	0
प	99.8	0.2	0
फ	99.8	0.2	0
ब	99.8	0.2	0
भ	99.8	0.2	0
म	99.8	0.2	0
य	99.8	0.2	0
र	99.8	0.2	0
ल	99.8	0.2	0
ळ	99.8	0.2	0
व	99.8	0.2	0
ह	99.8	0.2	0
श	99.8	0.2	0
ष	99.8	0.2	0
स	99.8	0.2	0

resource centers across India. The notable applications being developed include Hindi spell checkers, Hindi mono and multi lingual lexicons, Hindi text to speech systems and Hindi speech recognition systems. The work is also being done on Hindi machine translations systems with English and Indian languages, Hindi OCR systems and other related applications and resources including corpora, POS taggers, morphological analyzers, etc. Information about this work is available at [8]. The Hindi WordNet has also been released [8, 36] (Table 8.7).

Finally we conclude the Chapter with a note that we are exploring further results on HP Labs India Indic Handwriting dataset using other soft computing techniques such as fuzzy and rough version of hierarchical bidirectional recurrent neural networks.

References

1. Bajaj, R., Dey, L., Chaudhury, S., Devnagari Numeral Recognition by combining Decision of Multiple Connectionist Classifiers, Sadhana, 27(1), pp 59–72, 2002.
2. Bansal, V., Sinha, R. M. K., Integrating Knowledge Sources in Devanagari Text Recognition System, IEEE Transactions on Systems, Man, and Cybernetics - Part A: Systems and Humans, 30(4), pp 500–505, 2000.
3. Bunke, H., Wang, P. S. P. (Editors), Handbook of Character Recognition and Document Image Analysis, World Scientific, 1997.
4. Chaudhuri, A., Fuzzy Rough Support Vector Machine for Data Classification, International Journal of Fuzzy System Applications, 5(2), pp 26–53, 2016.
5. Chaudhuri, A., Modified Fuzzy Support Vector Machine for Credit Approval Classification, AI Communications, 27(2), pp 189–211, 2014.
6. Chaudhuri, A., De, Fuzzy Support Vector Machine for Bankruptcy Prediction, Applied Soft Computing, 11(2), pp 2472–2486, 2011.
7. Chaudhuri, A., Applications of Support Vector Machines in Engineering and Science, Technical Report, Birla Institute of Technology Mesra, Patna Campus, India, 2011.
8. Chaudhuri, A., Some Experiments on Optical Character Recognition Systems for different Languages using Soft Computing Techniques, Technical Report, Birla Institute of Technology Mesra, Patna Campus, India, 2010.
9. Chaudhuri, A., De, K., Job Scheduling using Rough Fuzzy Multi-Layer Perception Networks, Journal of Artificial Intelligence: Theory and Applications, 1(1), pp 4–19, 2010.
10. Chaudhuri, A., De, K., Chatterjee, D., Discovering Stock Price Prediction Rules of Bombay Stock Exchange using Rough Fuzzy Multi-Layer Perception Networks, Book Chapter: Forecasting Financial Markets in India, Rudra P. Pradhan, Indian Institute of Technology Kharagpur, (Editor), Allied Publishers, India, pp 69–96, 2009.
11. Cheriet, M., Kharma, N., Liu, C. L., Suen, C. Y., Character Recognition Systems: A Guide for Students and Practitioners, John Wiley and Sons, 2007.
12. De, R. K., Basak, J., Pal, S. K., Neuro-Fuzzy Feature Evaluation with Theoretical Analysis, Neural Networks, 12(10), pp 1429–1455, 1999.
13. De, R. K., Pal, N. R., Pal, S. K., Feature Analysis: Neural Network and Fuzzy Set Theoretic Approaches, Pattern Recognition, 30(10), pp 1579–1590, 1997.
14. Gonzalez, R. C., Woods, R. E., Digital Image Processing, 3rd Edition, Pearson, 2013.
15. Jain, A. K., Fundamentals of Digital Image Processing, Prentice Hall, India, 2006.
16. Kompalli, S., Setlur, S., Design and Comparison of Segmentation driven and Recognition driven Devanagari OCR, International Conference on Document Image Analysis for Libraries, pp 96–102, 2006.
17. Pal, S. K., Mitra, S., Mitra, P., Rough-Fuzzy Multilayer Perception: Modular Evolution, Rule Generation and Evaluation, IEEE Transactions on Knowledge and Data Engineering, 15(1), pp 14–25, 2003.
18. Pal, U., Sharma, N., Wakabayashi, T., Kimura, F., Offline Handwritten Character Recognition of Devnagari Script, International Conference on Document Analysis and Recognition, pp 496–500, 2007.
19. Pal, U., Chaudhuri, B. B., Printed Devnagari script OCR system, Vivek, 10(1), pp 12–24, 1997.

20. Russ, J. C., The Image Processing Handbook, CRC Press, 6th Edition, 2011.
21. Sethi, K. Chatterjee, B., Machine Recognition of Constrained Hand Printed Devnagari, Pattern Recognition, 9(2), pp 69–77, 1977.
22. Sharma, N., Pal, U., Kimura, F., Pal, S., Recognition of Offline Handwritten Devnagari Characters using Quadratic Classifier, Indian Conference on Computer Vision, Graphics and Image Processing, pp 805–816, 2006.
23. Taghva, K., Borsack, J., Condit, A., Effects of OCR Errors on Ranking and Feedback using the Vector Space Model, Information Processing and Management, 32(3), pp 317–327, 1996.
24. Taghva, K., Borsack, J., Condit, A., Evaluation of Model Based Retrieval Effectiveness with OCR Text, ACM Transactions on Information Systems, 14(1), pp 64–93, 1996.
25. Taghva, K., Borsack, J., Condit, A., Erva, S., The Effects of Noisy Data on Text Retrieval, Journal of American Society for Information Science, 45 (1), pp 50–58, 1994.
26. Young, T. Y., Fu, K. S., Handbook of Pattern Recognition and Image Processing, Academic Press, 1986.
27. Zadeh, L. A., Fuzzy Sets, Information and Control, 8(3), pp 338–353, 1965.
28. Zeng, J., Liu, Z. Q., Type-2 Fuzzy Markov Random Fields and their Application to Handwritten Chinese Character Recognition, IEEE Transactions on Fuzzy Systems, 16(3), pp 747–760, 2008.
29. Zimmermann, H. J., Fuzzy Set Theory and its Applications, 4th Edition, Kluwer Academic Publishers, Boston, 2001.
30. https://en.wikipedia.org/wiki/Hindi.
31. http://lipitk.sourceforge.net/datasets/dvngchardata.htm.
32. http://www.i2ocr.com/free-online-hindi-ocr.
33. http://homepages.cwi.nl/~dik/english/codes/indic.html.
34. http://varamozhi.sourceforge.net/iscii91.pdf.
35. http://acharya.iitm.ac.in/multi_sys/exist_codes.html.
36. http://tdil.mit.gov.in/pchangeuni.htm.
37. http://www.iiit.net/ltrc/FC-1.0/fc.html.
38. http://scripts.sil.org/cms/scripts/page.php?site_id=nrsi&id=EncCnvtrs.

Chapter 9
Optical Character Recognition Systems for Gujrati Language

Abstract The optical character recognition (OCR) systems for Gujrati language were the most primitive ones and occupy a significant place in pattern recognition. The Gujrati language OCR systems have been used successfully in a wide array of commercial applications. The different challenges involved in the OCR systems for Gujrati language is investigated in this Chapter. The pre-processing activities such as binarization, noise removal, skew detection, character segmentation and thinning performed on the datasets considered. The feature extraction is performed through fuzzy Genetic Algorithms (GA). The feature based classification is performed through important soft computing techniques viz rough fuzzy multilayer perceptron (RFMLP), fuzzy support vector machine (FSVM), fuzzy rough support vector machine (FRSVM) and fuzzy markov random fields (FMRF). The superiority of soft computing techniques is demonstrated through the experimental results.

Keywords Gujrati language OCR · RFMLP · FSVM · FRSVM · FMRF

9.1 Introduction

Several optical character recognition (OCR) systems have been developed for Gujrati language [1]. Gujrati is the most widely spoken language across the Indian subcontinent [2] and several other parts of the world after Mandarin, Spanish and English. This motivation has led to the development of OCR systems for Gujrati language [3]. Gujrati language OCR systems have been used successfully in a wide array of commercial applications [1]. The character recognition of Gujrati language has a high potential in data and word processing among other Indian languages. Some commonly used applications of the OCR systems of Gujrati language [1] are automated postal address and ZIP code reading, data acquisition in bank checks, processing of archived institutional records etc. as evident in other languages such as English, French, German and Latin.

© Springer International Publishing AG 2017   217
A. Chaudhuri et al., *Optical Character Recognition Systems for Different Languages with Soft Computing*, Studies in Fuzziness and Soft Computing 352,
DOI 10.1007/978-3-319-50252-6_9

The first complete OCR in Gujrati language was introduced by [4] where structural and template features are used for recognizing basic, modified and compound characters. To recognize real life printed character documents of varying size and font [5] have proposed statistical features. [6] has used density, moment of curve and descriptive component for recognizing Gujrati handwritten characters and numerals. A set of primitives [1], such as global and local horizontal and vertical line segments, right and left slant and loop for recognizing handwritten Gujrati characters have been proposed earlier. Also directional chain code information [1] of the contour points of the characters for recognizing handwritten Gujrati characters have been used.

The standardization of OCR character set for Gujrati language was provided through ISCII (IS 13194:1988) [2] as shown in the Fig. 9.1. It is the national standard for Devanagari character set encoding based on earlier standard IS 10402:1982 [1]. ISCII is a standard for Devanagari script and has been used for other Indian languages. The standard contains ASCII in lower 128 slots and Devanagari alphabet superset in upper 128 slots and therefore it is a single byte standard. Though it is primarily an encoding standard and sorting is usually not catered directly in such standards, the standard was devised to do some implicit sorting directly on encoding. The variations of ISCII include PC-ISCII and language specific ISCII charts [7]. The unicode provides an international standard for Devanagari character set encoding based on IS 13194:1988 from 0900 till 097F and therefore is not exactly equivalent to IS 13194:1991 [8, 9]. This has been used for Gujrati and other Devanagari script based languages including Marathi, Sanskrit, Prakrit, Sindhi, etc. There are other encodings which have been used by vendors which not prevalent anymore. The Gujrati language based OCR systems is thus one of the most fascinating and challenging areas of Indian pattern recognition research community [1].

Gujarati

◌̐ 0A81	◌̇ 0A82	◌ઃ 0A83	અ 0A85	આ 0A86	ઇ 0A87	ઈ 0A88	ઉ 0A89	ઊ 0A8A	ઋ 0A8B	ઌ 0A8C
ઍ 0A8D	એ 0A8F	ઐ 0A90	ઑ 0A91	ઓ 0A93	ઔ 0A94	ક 0A95	ખ 0A96	ગ 0A97	ઘ 0A98	ઙ 0A99
ચ 0A9A	છ 0A9B	જ 0A9C	ઝ 0A9D	ઞ 0A9E	ટ 0A9F	ઠ 0AA0	ડ 0AA1	ઢ 0AA2	ણ 0AA3	ત 0AA4
થ 0AA5	દ 0AA6	ધ 0AA7	ન 0AA8	પ 0AAA	ફ 0AAB	બ 0AAC	ભ 0AAD	મ 0AAE	ય 0AAF	ર 0AB0
લ 0AB2	ળ 0AB3	વ 0AB5	શ 0AB6	ષ 0AB7	સ 0AB8	હ 0AB9	◌઼ 0ABC	ઽ 0ABD	◌ા 0ABE	◌િ 0ABF
◌ી 0AC0	◌ુ 0AC1	◌ૂ 0AC2	◌ૃ 0AC3	◌ૄ 0AC4	◌ે 0AC5	◌ૈ 0AC7	◌ૉ 0AC8	◌ો 0AC9	◌ૌ 0ACB	◌ૂ 0ACC
◌્ 0ACD	ૐ 0AD0	ૠ 0AE0	ૡ 0AE1	◌ૢ 0AE2	◌ૣ 0AE3	૦ 0AE6	૧ 0AE7	૨ 0AE8	૩ 0AE9	૪ 0AEA
૫ 0AEB	૬ 0AEC	૭ 0AED	૮ 0AEE	૯ 0AEF	▢ 0AF0	૱ 0AF1				

Fig. 9.1 The ISCII code chart for Gujrati characters

The OCR process for any language involves extraction of defined characteristics called features to classify an unknown character into one of the known classes [1, 10, 11] to a user defined accuracy level. As such any good OCR system is best defined in terms of the success of feature extraction and classification tasks. The same is true for Gujrati language. The process becomes tedious in case the language has dependencies where some characters look identical. Thus the classification becomes a big challenge.

In this chapter we start the investigation of OCR systems considering the different challenges involved in Gujrati language. The different pre-processing activities such as binarization, noise removal, skew detection and correction, character segmentation and thinning are performed on the considered datasets [12]. The feature extraction is performed through discrete Hough transformation. The feature based classification is performed through important soft computing techniques viz rough fuzzy multilayer perceptron (RFMLP) [1, 13–16] two support vector machine (SVM) based methods such as fuzzy support vector machine (FSVM) [17, 18] and fuzzy rough support vector machine (FRSVM) [19, 20] and fuzzy markov random fields (FMRF) [21]. The experimental results demonstrate the superiority of soft computing techniques over the traditional methods.

This chapter is structured as follows. In Sect. 9.2 a brief discussion about the Gujrati language script and datasets used for experiments are presented. The different challenges of OCR for Gujrati language are highlighted in Sect. 9.3. The next section illustrates the data acquisition. In Sect. 9.5 different pre-processing activities on the datasets such as binarization, noise removal, skew detection and correction, character segmentation and thinning are presented. This is followed by a discussion of feature extraction on Gujrati language dataset in Sect. 9.6. The Sect. 9.7 explains the state of art of OCR for Gujrati language in terms of feature based classification methods. The corresponding experimental results are given in Sect. 9.8. Finally Sect. 9.9 concludes the Chapter with some discussions and future research directions.

9.2 Gujrati Language Script and Experimental Dataset

In this section we present brief information about the Gujrati language script and the dataset used for experiments. Gujrati language spelled as ગુજરાતી [2] is an Indo Aryan language native to the Indian state of Gujrat. It is part of the greater Indo-European language family. Gujrati is descended from Old Gujarati (circa 1100–1500 AD). It is the official language in the state of Gujrat as well as an official language in the union territories of Daman and Diu and Dadra and Nagar Haveli. The Gujrati script is adapted from the Devanagari script to write the Gujrati language. The earliest known document in the Gujrati script is a manuscript dating from 1592, and the script first appeared in print in a 1797 advertisement. Until the 19th century it was used mainly for writing letters and keeping accounts, while the Devanagari script was used for literature and academic writings. The Gujrati script is also known as the banker's script. It is a syllabic alphabet in that consonants all have an inherent vowel. Vowels can be written as independent letters, or by using a variety of diacritical marks which are written above, below, before or after the consonant they belong to. The Gujrati language consists of 28 vowels and 36 consonants as shown in Fig. 9.2 [2]. It is written from left to right.

According to the Central Intelligence Agency, 4.5% of the Indian population (1.21 billion according to the 2011 census) speak Gujrati which amounts to 54.6 million speakers in India. There are about 65.5 million speakers of Gujrati worldwide.

Of the approximately 46 million speakers of Gujrati in 1997, roughly 45.5 million resided in India, 150,000 in Uganda, 50,000 in Tanzania, 50,000 in Kenya and roughly 100,000 in Karachi, Pakistan. There is a certain amount of Mauritian population and a large amount of Réunion Island people who are from Gujrati descent among which some of them still speak Gujrati.

A considerable Gujrati speaking population exists in North America most particularly in the New York City Metropolitan Area and in the Greater Toronto Area and also in the major metropolitan areas of the United States and Canada. The UK has 200,000 Gujrati speakers many of them situated in London area and also in Birmingham, Manchester Warrington and in Leicester, Coventry, Bradford and the former mill towns within Lancashire and Wembley. Besides being spoken by the Gujrati people, non-Gujrati residents of and migrants to the state of Gujarat also count as speakers. Among them are Kutchis, Parsis and Hindu Sindhi refugees from Pakistan.

The Gujrati language dataset used for performing OCR experiments is the Gujrati Character Recognition dataset and is adapted from [12]. The database contains Gujrati text which is used here to for training and testing. The database contains unconstrained handwritten text which are scanned at a resolution of 300 dpi and saved as PNG images with 256 gray levels. The Fig. 9.3 shows a sample snapshot from the database. Further details are available at [12].

9.3 Challenges of Optical Character Recognition Systems for Gujrati Language

The OCR for Gujrati language has become one of the most successful applications in pattern recognition and artificial intelligence among the Indian languages after Bengali language [1, 3, 10]. It is rigorously perused after investigation topic by the Indian OCR research community [1]. The most commercially available

Fig. 9.2 The Gujrati language consisting of 28 vowels and 36 consonants

Fig. 9.3 The Gujrati character recognition dataset

અ આ ઇ ઈ ઉ ઊ ઋ

એ ઐ ઓ ઔ

ક ખ ગ ઘ ઙ

ચ છ જ ઝ ઞ

ટ ઠ ડ ઢ ણ

ત થ દ ધ ન

પ ફ બ ભ મ

ય ર લ વ

શ ષ સ હ ળ

OCR system for Gujrati language is Gujarati OCR [4]. Considering the important aspects of versatility, robustness and efficiency, the commercial OCR systems for Gujrati language are generally divided into three generations [1]. The methods mentioned in Sect. 9.1 do not consider shape variation for extracting features. But in most of the Indian languages especially Gujrati, a large number of similar shape type characters such as basic and conjunct are present. From this point of view, novel features based on the topography of a character has to be used to improve the performance of existing OCR in Gujrati script. The major features worth mentioning are as follows:

(a) The main challenge to design an OCR for Gujrati language is to handle large scale shape variation among different characters. The strokes in characters can be decomposed into segments which are straight lines, convexities or closed boundaries or hole. The topography of character strokes from 4 viewing directions and different convex shapes formed by the character strokes with respect to the presence of closed region boundaries need to be considered.

(b) The extracted features are represented by a shape based graph where each node contains the topographic feature and they all are placed with respect to their centroids and relative positions in the original character image.

(c) This topographic feature set helps to differentiate very similar shape and type characters in a proper way.

Despite some rigorous research and existence of established commercial OCR products based on Gujrati language, the output from such OCR processes often contains errors. The more highly degraded is input, the greater is error rate. Since inputs form the first stage in a pipeline where later stages are designed to support sophisticated information extraction and exploitation applications, it is important to understand the effects of recognition errors on downstream analysis routines. Few questions are required to be addressed in this direction. They are as follows:

(a) Are all recognition errors equal in impact or some are worse than others?
(b) Can the performance of each stage be optimized in isolation or the end-to-end system should be considered?
(c) In balancing the trade-off between the risk of over and under segmenting characters during OCR where should the line be drawn to maximize overall performance?

The answers to these questions often influence the way OCR systems for Gujrati language are designed and build for analysis [10].

The Gujrati language OCR system converts numerous published books in Gujrati language into editable computer text files. The latest research in this area has grown to incorporate some new methodologies to overcome the complexity of Gujrati writing style. All these algorithms have still not been tested for complete characters of Gujrati alphabet. Hence, there is a quest for developing an OCR system which handles all classes of Gujrati text and identify characters among these classes increasing versatility, robustness and efficiency in commercial OCR systems. The recognition of printed Gujrati characters is itself a challenging problem since there is a variation of the same character due to change of fonts or introduction of different types of noises. There may be noise pixels that are introduced due to scanning of the image. A significant amount of research has been done towards text data processing in Gujrati language from noisy sources [3]. The majority of the work has focused predominately on errors that arise during speech recognition systems [10]. Several research papers have appeared which examines the noise problem from variety of perspectives with most emphasizing issues that are inherent in written and spoken Gujrati language [1]. However, there has been less work concentrating on noise induced by OCR. Some earlier works by [22] show that moderate error rates have little impact on effectiveness of traditional information retrieval measures. However, this conclusion is tied to certain assumptions about information retrieval through bag of words, OCR error rate which may not be too high and length of documents which may not be too short. Some other notable research works in this direction are given in [23, 24]. All these works try to address some significant issues involved in OCR systems for Gujrati language such as error correction, performance evaluation etc. involving flexible and rigorous mathematical treatment [1]. Besides this any Gujrati character can be represented in variety of fonts and sizes as per the needs and requirements of application. Further the character with same font and size may also be bold face character as well as normal one [10]. Thus the width of stroke also significantly affects recognition process. Therefore, a good character recognition approach for Gujrati language [1, 25–27]:

(a) Must eliminate noise after reading binary image data
(b) Smooth image for better recognition
(c) Extract features efficiently
(d) Train the system and
(e) Classify patterns accordingly.

9.4 Data Acquisition

The progress in automatic character recognition systems in Gujrati language is generally evolved in two categories according to the mode of data acquisition which can be either online or offline character recognition systems. Offline character recognition captures data from paper through optical scanners or cameras whereas online recognition systems utilize digitizers which directly capture writing with the order of strokes, speed, pen-up and pen-down information. As such the scope of this text is restricted to OCR systems, we confine our discussion to offline character recognition [3] for Gujrati language. The Gujrati language datasets used in this research is mentioned in Sect. 9.2.

9.5 Data Pre-processing

Once the data has been acquired properly we proceed to pre-process the data. In pre-processing stage [3] a series of operations are performed viz binarization, noise removal, skew detection, character segmentation and thinning or skeletonization. The main objective of pre-processing is to organize information so that the subsequent character recognition task becomes simpler. It essentially enhances the image rendering it suitable for segmentation.

9.5.1 Binarization

Binarization [19] is an important first step in character recognition process. A large number of binarization techniques are available in the literature [6] each of which is appropriate to particular image types. Its goal is to reduce the amount of information present in the image and keep only the relevant information. Generally the binarization techniques of gray scale images are classified into two categories viz overall threshold where single threshold is used in the entire image to form two classes (text and background) and local threshold where values of thresholds are determined locally (pixel-by-pixel or region-by-region). Here, we calculate threshold Th of each pixel locally by [3]:

$$Th = (1 - k) \cdot m + k \cdot m + k \cdot \frac{\sigma}{R(m - M)} \qquad (9.1)$$

In Eq. (9.1) k is set to 0.5; m is the average of all pixels in window; M is the minimum image grey level; σ is the standard deviation of all pixels in window and R is the maximum deviation of grayscale on all windows.

9.5.2 Noise Removal

The scanned text documents often contain noise that arises due to printer, scanner, print quality, document age etc. Therefore, it is necessary to filter noise [6] before the image is processed. Here a low-pass filter is used to process the image [3] which is used for later processing. The main objective in the design of a noise filter is that it should remove as much noise as possible while retaining the entire signal [19].

9.5.3 Skew Detection and Correction

When a text document is fed into scanner either mechanically or manually a few degrees of tilt or skew is unavoidable. In skew angle the text lines in digital image make angle with horizontal direction. A number of methods are available in literature for identifying image skew angles [6]. They are basically categorized on the basis of projection profile analysis, nearest neighbor clustering, Hough transform, cross correlation and morphological transforms. The aforementioned methods correct the detected skew. Some of the notable research works in this direction are available in [3]. Here Hough transform is used for skew detection and correction [3].

9.5.4 Character Segmentation

Once the text image is binarized, noise removed and skew corrected, the actual text content is extracted. This process leads to character segmentation [28]. The commonly used segmentation algorithms in this direction are connected component labeling, x-y tree decomposition, run length smearing and Hough transform [6]. Here Hough transform is used for character segmentation [3].

9.5.5 Thinning

The character segmentation process is followed by thinning or skeletonization [29]. In thinning one-pixel-width representation or skeleton of an object is obtained by preserving connectedness of the object and its end points [18]. The thinning process reduces image components to their essential information so that further analysis and recognition are facilitated. For instance, an alphabet can be handwritten with different pens giving different stroke thicknesses but information presented is same. This enables easier subsequent detection of pertinent features. As an illustration consider letter d shown in Fig. 9.4 before and after thinning. A

Fig. 9.4 An image before
and after thinning

number of thinning algorithms have been used in the past with considerable success. The most common algorithm used is the classical hilditch algorithm and its
variants [3]. Here hilditch algorithm is used for thinning [19]. For recognizing
large graphical objects with filled regions which are often found in logos, region
boundary detection is useful but for small regions corresponding to individual
characters neither thinning nor boundary detection is performed. Rather entire
pixel array representing the region is forwarded to subsequent stage of analysis.

9.6 Feature Selection Through Genetic Algorithms

As mentioned in Chap. 4 the heart of any OCR system is the formation of feature vector used in recognition stage. This fact is also valid for Gujrati language
OCR system. This phase extracts the features from segmented areas of image containing characters to be recognized that distinguishes an area corresponding to
a letter from an area corresponding to other letters. The feature extraction phase
can thus be visualised as finding a set of parameters or features that define the
shape of character as precise and unique. In Chap. 3 the term feature extraction
is often used synonymously by feature selection which refers to algorithms that
select the best subset of input feature set. These methods create new features based
on transformations or combination of original features [20]. The features selected
help in discriminating the characters. Achieving high recognition performance is
attributed towards the selection of appropriate feature extraction methods. A large
number of OCR based feature extraction methods are available in literature [6, 19]
except that the selected method depends on the application concerned. There is no
universally accepted set of feature vectors in OCR. The features that capture topological and geometrical shape information are the most desired ones. The features
that capture spatial distribution of black (text) pixels are also very important [6].
The genetic algorithm based feature extraction approach based on multilayer perceptron (MLP) is successfully applied for the OCR of Gujrati language [1].

A number of neural network and fuzzy set theoretic approaches have been proposed for feature analysis in recent past [5]. A feature quality index (FQI) measure
for ranking of features has been suggested by [17]. The feature ranking process is
based on influence of feature on MLP output. It is related to the importance of feature in discriminating among classes. The impact of qth feature on MLP output out
of a total of p features is measured by setting feature value to zero for each input
pattern $x_i, i = 1, \ldots, n$. FQI is defined as the deviation of MLP output with qth
feature value set to zero from output with all features present such that:

$$FQI_q = \frac{1}{n} \sum_{i=1}^{n} \left\| OV_i - OV_i^{(q)} \right\|^2 \tag{9.2}$$

In Eq. (9.2) OV_i and $OV_i^{(q)}$ are output vectors with all p features present and with qth feature set to zero. The features are ranked according to their importance as $q_1, \ldots, q_p$ if $FQI_{q_1} > \cdots > FQI_{q_p}$. In order to select best p' features from the set of p features, $\binom{p}{p'}$ possible subsets are tested one at a time. The quality index $FQI_k^{(p')}$ of kth subset S_k is measured as:

$$FQI_k^{(p')} = \frac{1}{n} \sum_{i=1}^{n} \left\| OV_i - OV_i^k \right\|^2 \tag{9.3}$$

In Eq. (9.3) OV_i^k is MLP output vector with x_i^k as input where x_i^k is derived from x_i as:

$$x_{ij}^k = \begin{cases} 0 & \text{if } j \in S_k \\ x_{ij} & \text{ow} \end{cases} \tag{9.4}$$

A subset S_j is selected as optimal set of features if $FQI_j^{(p')} \geq FQI_k^{(p')} \ \forall k, k \neq j$. An important observation here is that value of p' should be predetermined and $\binom{p}{p'}$ number of possible choices are verified to arrive at the best feature set. It is evident that no a priori knowledge is usually available to select the value p' and an exhaustive search is to be made for all values p' with $p' = 1, \ldots, p$. The number of possible trials is $(2^P - 1)$ which is prohibitively large for high values of p. To overcome drawbacks of above method, best feature set is selected by using of genetic algorithms [18]. Let us consider mask vector M where $M_i \in \{0, 1\}; i = 1, \ldots, p$ and each feature element $q_i, i = 1, \ldots, n$ is multiplied by corresponding mask vector element before reaching MLP input such that $I_i = q_i M_i$. MLP inputs are then written as:

$$I_i = \begin{cases} 0 & \text{if } M_i = 0 \\ q_i & \text{ow} \end{cases} \tag{9.5}$$

Thus, a particular feature q_i reaches MLP if corresponding mask element is one. To find sensitivity of a particular feature q_j, mask bit M_j is set to zero. With respect to the above discussions when kth subset of feature set $\{q_1, \ldots, q_p\}$ is selected, all corresponding mask bits are set to zero and rest are set to one. When feature set multiplied by these mask bits reaches MLP, the effect of setting features of subset S_k to zero is obtained. Then value of FQI_k is calculated. It is to be noted that kth. subset thus chosen may contain any number of feature elements and not pre-specified p' number of elements. Starting with an initial population of strings representing mask vectors, genetic algorithm is used with reproduction, crossover and mutation operators to determine best value of objective function. The objective function is FQI value of feature set S_k selected with mask bits set to zero for specific features and is given by:

$$FQI_k = \frac{1}{n} \sum_{i=1}^{n} \left\| OV_i - OV_i^k \right\|^2 \tag{9.6}$$

In this process, both the problems of predetermining value of p' and searching through $\binom{p}{p'}$ possible combinations for each value of p'. In genetic algorithm implementation, the process is started with 20 features generated from fuzzy Hough transform such that the number of elements in mask vector is also 20. After running genetic algorithm for sufficiently large number of generations, mask string with best objective function value is determined. The feature elements corresponding to mask bits zero are chosen as selected set of features. The parameters for genetic algorithms are determined in terms of chromosome length, population size, mutation probability and crossover probability. MLP is next trained with only selected feature set for classification. The number of features varied when required. The genetic algorithm based feature selection method is shown in Fig. 9.5.

9.7 Feature Based Classification: Sate of Art

After extracting the essential features from the pre-processed character image the concentration pointer turns on the feature based classification methods for OCR of Gujrati language. Considering the wide array of feature based classification methods for OCR, these methods are generally grouped into following four broad categories [17, 28, 30]:

(a) Statistical methods
(b) ANN based methods
(c) Kernel based methods
(d) Multiple classifier combination methods.

We discuss here the work done on Gujrati characters using some of the above-mentioned classification methods. All the classification methods used here are soft computing based techniques. The next three subsections highlight the feature based classification based on RFMLP [1, 13, 14, 16], FSVM [17, 18], FRSVM [19, 20] and FMRF [21] techniques. These methods are already discussed in Chap. 3. For further details interested readers can refer [1].

Fig. 9.5 The feature
selection process using
genetic algorithm

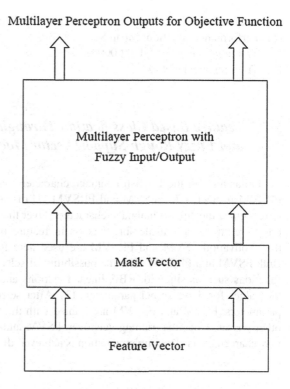

Multilayer Perceptron Outputs for Objective Function

9.7.1 Feature Based Classification Through Rough Fuzzy Multilayer Perceptron

The Gujrati language characters are recognized here through RFMLP [3] which is discussed in Chap. 3. For the Gujrati language RFMLP has been used with considerable success [1]. The diagonal feature extraction scheme is used for drawing of features from the characters. For classification stage we use these features extracted. RFMLP used here is a feed forward network with back propagation ANN having four hidden layers. The architecture used is 70–100–100–100–100–100–52 [3] for classification. The four hidden layers and output layer uses tangent sigmoid as the activation function. The feature vector is again denoted by $F = (f_1, \ldots, f_m)$ where m denotes number of training images and each f has a length of 70 which represent the number of input nodes. The 64 neurons in the output layer correspond to 28 vowels and 36 consonants of Gujrati alphabets. The network training parameters are briefly summarized as [3]:

(a) Input nodes: 70
(b) Hidden nodes: 100 at each layer
(c) Output nodes: 64

(d) Training algorithm: Scaled conjugate gradient backpropagation
(e) Performance function: Mean Square Error
(f) Training goal achieved: 0.000002
(g) Training epochs: 4000.

9.7.2 Feature Based Classification Through Fuzzy and Fuzzy Rough Support Vector Machines

In similar lines to the English language character recognition through FSVM and FRSVM in Sect. 4.7.3, FSVM and FRSVM [1] discussed in Chap. 3 is used here to recognize the Gujrati language characters. Over the past years SVM based methods have shown a considerable success in feature based classification [5]. With this motivation FSVM and FRSVM are used here for feature classification task. Both FSVM and FRSVM offer the possibility of selecting different types of kernel functions such as sigmoid, RBF, linear functions and determining the best possible values for these kernel parameters [3]. After selecting the kernel type and its parameters, FSVM and FRSVM are trained with the set of features obtained from other phases. Once the training gets over, FSVM and FRSVM are used to classify new character sets. The implementation is achieved through LibSVM library [20].

9.7.3 Feature Based Classification Through Fuzzy Markov Random Fields

After RFMLP, FSVM and FRSVM, the Gujrati language characters are recognized through FMRFs [21] which is discussed in Chap. 3. For the Gujrati language FMRFs has been used with considerable success [1]. FMRFs are born by integrating type-2 (T2) fuzzy sets with markov random fields (MRFs) resulting in T2 FMRFs. The T2 FMRFs handles both fuzziness and randomness in the structural pattern representation. The T2 membership function has a 3-D structure in which the primary membership function describes randomness and the secondary membership function evaluates the fuzziness of the primary membership function [31, 32]. The MRFs represent statistical patterns structurally in terms of neighborhood system and clique potentials [21]. The T2 FMRFs defines the same neighborhood system as the classical MRFs. In order to describe the uncertain structural information in patterns, the fuzzy likelihood clique potentials are derived from T2 fuzzy Gaussian mixture models. The fuzzy prior clique potentials are penalties for the mismatched structures based on prior knowledge. The T2 FMRFs models the hierarchical character structures present in the language characters. For interested readers further details are available in [1]. The evaluation is done on different types of highly biased character data. The implementation of FMRFs is performed in MATLAB [1].

9.8 Experimental Results

In this section, the experimental results for soft computing tools viz RFMLP, FSVM, FRSVM and FMRF on the Gujrati language dataset [12] highlighted in Sect. 9.2 are presented. The prima face is to select the best OCR system for Gujrati language [3].

9.8.1 Rough Fuzzy Multilayer Perceptron

The experimental results for a subset of the Gujrati characters both vowels and consonants using RFMLP are shown in Table 9.1. Like the Hindi characters, the Gujrati characters are categorized as ascenders (characters such as પ, આ, ઉ, ઊ etc.) and descenders (characters such as ય, ઘ, ૫, મ, ષ etc.). RFMLP shows better results than the traditional methods [1]. The testing algorithm is applied to all standard cases of characters [12] in various samples. An accuracy of around 99% has been achieved in all the cases. However, after successful training and testing of algorithm the following flaws are encountered [1]:

(a) There may an instance when there is an occurrence of large and dispro-portionate symmetry in both ascenders as well as descenders as shown in Fig. 9.6a, b.
(b) There may be certain slants in certain characters which results in incorrect detection of characters as shown in Fig. 9.7.
(c) Sometimes one side of the image may have less white space and more pixel concentration whereas other side may have more white space and less pixel concentration. Due to this some characters are wrongly detected as shown in Fig. 9.8.

9.8.2 Fuzzy and Fuzzy Rough Support Vector Machines

FSVM and FRSVM show promising results for feature classification task through different types of kernel functions by selecting the best possible values for kernel parameters [17, 18, 20]. For testing accuracy of the system in the first test case sce-nario we use an image which contained 100 letters as shown in Fig. 9.9. The train-ing set is constructed through two images containing 40 examples of each vowel letter in the Gujrati alphabet which took around 19.89 s. The results are presented in Tables 9.2 and 9.3. The parameter C is regularization parameter, ρ is bias term, κ. is kernel function, σ is smoothing function which reduces variance and ϕ is mapping function in feature space. Further discussion on these parameters is available in [1].

For the final test case scenario we use for training the features corresponding to Gujrati letters. The images used for testing are the ones used in the first and

Table 9.1 The experimental results for a subset of the Gujrati characters using RFMLP

Characters	Successful Recognition (%)	Unsuccessful Recognition (%)	No Recognition (%)
ખ	99	1	0
આ	99	1	0
ઇ	99	1	0
ઈ	99	1	0
ઉ	99	1	0
ઊ	99	1	0
એ	98	1	1
ઐ	99	1	0
ઓ	99	1	0
ઔ	99	1	0
ઋ	99	1	0
ઍ	99	1	0
ક	98	1	1
ગ	98	1	1
ધ	99	1	0
ડ	99	1	0
ય	99	1	0
છ	99	1	0
જ	99	1	0
ઝ	99	1	0
ગ	99	1	0
ચ	99	1	0
શ	98	1	1
ટ	98	1	1
ઠ	99	1	0
ડ	99	1	0
ઢ	99	1	0
ણ	99	1	0
૨	99	1	0
ષ	99	1	0
ત	99	1	0
થ	99	1	0
૬	99	1	0
ધ	99	1	0
ન	99	1	0
લ	99	1	0
સ	99	1	0
પ	99	1	0
ફ	99	1	0
ળ	98	1	1
ભ	98	1	1
મ	99	1	0
વ	99	1	0

(a)   **(b)**

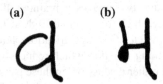

Fig. 9.6 The disproportionate symmetry in characters 'વ' and 'મ'

Fig. 9.7 The character 'પ' with slant

Fig. 9.8 The uneven pixel combination for character 'પ'; region-i on the *left image* has less white space than the *right image*

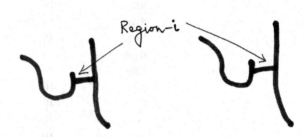

Region-i

Fig. 9.9 The test image subset for Gujrati letters

ખ ખખખ આઆઆ આ

ઘ ઘઘઘ ઇ ઇ ઇ ઇ

ઉ ઉઉઉ ૩ ૩ ૩ ૩

એ એએએ એ એએએએ

ક ક કક અ અઅઅ

ધ ધ ધ ધ ડ ડ ડ ડ

ચ ચચચ ટ ટ ટ ટ

ષ ષ ષ ષ તતતત

પ પ પ પ વ વ વ વ

સ સ સસ ડ ડ ડ ડ

Table 9.2 The training set results corresponding to Gujrati letters (for FSVM)

Kernel Function	C	ρ	κ	σ	ϕ	Precision (%)
Linear	1	–	–	–	–	94
Linear	10	–	–	–	–	96
Linear	100	–	–	–	–	95
RBF	10	–	–	0.25	–	94
RBF	10	–	–	0.15	–	94
RBF	10	–	–	0.10	–	96
RBF	10	–	–	0.05	–	95
RBF	10	–	–	0.03	–	96
RBF	10	–	–	0.02	–	96
RBF	10	–	–	0.01	–	98
RBF	10	–	–	0.005	–	97
Polynomial	10	2	2	–	–	96
Polynomial	10	2	4	–	–	96
Polynomial	10	2	1	–	–	96
Polynomial	10	2	0.5	–	–	95
Polynomial	10	3	2	–	–	93
Polynomial	10	3	4	–	–	93
Polynomial	10	3	1	–	–	95
Polynomial	10	3	0.5	–	–	95
Polynomial	10	4	2	–	–	95
Polynomial	10	4	4	–	–	94
Polynomial	10	4	1	–	–	96
Polynomial	10	4	0.5	–	–	95
Sigmoid	10	–	0.5	–	1	93
Sigmoid	10	–	0.5	–	5	93
Sigmoid	10	–	0.2	–	1	94
Sigmoid	10	–	0.7	–	1	96

second test cases. The training set construction took around 37.98 s. The results are presented in Tables 9.4 and 9.5.

A comparative performance of the soft computing techniques (RFMLP, FSVM, FRSVM, FMRF) used for Gujrati language with the traditional techniques (MLP, SVM) is provided in Fig. 9.10 for samples of 30 datasets. It is to be noted that the architecture of MLP used for classification is 70–100–100–52 [20] a sigmoid kernel is used with SVM.

All the tests are conducted on PC having Intel P4 processor with 4.43 GHz, 512 MB DDR RAM @ 400 MHz with 512 kB che. The training set construction was the longest operation of the system where the processor was loaded to 25% and the application occupied around 54.16 MB of memory. During idle mode the application consumes 43.02 MB of memory.

Table 9.3 The training set results corresponding to Gujrati letters (for FRSVM)

Kernel function	C	ρ	κ	σ	ϕ	Precision (%)
Linear	1	–	–	–	–	94
Linear	10	–	–	–	–	96
Linear	100	–	–	–	–	93
RBF	10	–	–	0.25	–	95
RBF	10	–	–	0.15	–	95
RBF	10	–	–	0.10	–	96
RBF	10	–	–	0.05	–	96
RBF	10	–	–	0.03	–	97
RBF	10	–	–	0.02	–	97
RBF	10	–	–	0.01	–	98
RBF	10	–	–	0.005	–	98
Polynomial	10	2	2	–	–	98
Polynomial	10	2	4	–	–	98
Polynomial	10	2	1	–	–	98
Polynomial	10	2	0.5	–	–	95
Polynomial	10	3	2	–	–	95
Polynomial	10	3	4	–	–	93
Polynomial	10	3	1	–	–	95
Polynomial	10	3	0.5	–	–	95
Polynomial	10	4	2	–	–	95
Polynomial	10	4	4	–	–	94
Polynomial	10	4	1	–	–	95
Polynomial	10	4	0.5	–	–	95
Sigmoid	10	–	0.5	–	1	95
Sigmoid	10	–	0.5	–	5	96
Sigmoid	10	–	0.2	–	1	96
Sigmoid	10	–	0.7	–	1	97

9.8.3 Fuzzy Markov Random Fields

The experimental results for a subset of the Gujrati characters both vowels and consonants using FMRFs are shown in the Table 9.6. The Gujrati characters are categorized as ascenders (characters such as ય, ઘ, ૫, મ, ષ etc.) and descenders (characters such as પ, આ, ઉ, ઊ etc.) as illustrated in Sect. 9.8.1. FMRFs shows better results than the traditional methods shows better results than the traditional methods as well RFMLP [1]. The testing algorithm is applied to all standard cases of characters [12] in various samples. An accuracy of around 99.8% has been achieved in all the cases.

Table 9.4 The training set results corresponding to Gujrati letters (for FSVM)

Kernel function	C	ρ	κ	σ	ϕ	Precision (%)
Linear	1	–	–	–	–	86
Linear	10	–	–	–	–	89
RBF	10	–	–	0.25	–	93
RBF	10	–	–	0.10	–	93
RBF	10	–	–	0.05	–	86
RBF	10	–	–	0.01	–	89
Polynomial	10	2	2	–	–	88
Polynomial	10	3	2	–	–	89
Polynomial	10	4	2	–	–	94
Sigmoid	10	–	0.5	–	–	93
Sigmoid	10	–	0.2	–	–	96

Table 9.5 The training set results corresponding to Gujrati letters (for FRSVM)

Kernel function	C	ρ	κ	σ	ϕ	Precision (%)
Linear	1	–	–	–	–	87
Linear	10	–	–	–	–	93
RBF	10	–	–	0.25	–	93
RBF	10	–	–	0.10	–	96
RBF	10	–	–	0.05	–	89
RBF	10	–	–	0.01	–	89
Polynomial	10	2	2	–	–	89
Polynomial	10	3	2	–	–	94
Polynomial	10	4	2	–	–	93
Sigmoid	10	–	0.5	–	–	94
Sigmoid	10	–	0.2	–	–	96

9.9 Further Discussions

All the feature based classification based methods used in this Chapter give better results than the traditional approaches [20]. On comparing with other algorithms, it is observed that RFMLP work with high accuracy in almost all cases which include intersections of loop and instances of multiple crossings. This algorithm focussed on the processing of various asymmetries in the Gujrati characters. RFMLP function consider the conditions: (a) the validity of the algorithms for non-cursive Gujrati alphabets (b) the requirement of the algorithm that height of both upper and lower case characters to be proportionally same and (c) for extremely illegible handwriting the accuracy achieved by the algorithm is very less.

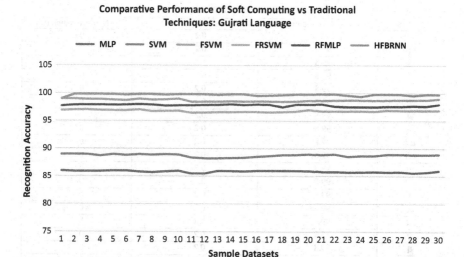

Fig. 9.10 The comparative performance of soft computing versus traditional techniques for Gujrati language

FSVM and FRSVM also give superior performance compared to traditional SVM [20]. FSVM and FRSVM achieve a precision rate up to 97%. in case of training with sets corresponding to either small or capital letters and up to 98% in case of training with sets corresponding to both small and capital letters respectively. Thus the system achieved its goal through the recognition of characters from an image. The future research direction entails in expanding the system through addition of techniques which determine automatically the optimal parameters of kernel functions. Further, any version of SVM can be used to better perform the feature based classification task. We are also experimenting with other robust versions of SVM which will improve the overall recognition accuracy of the Gujrati characters.

FMRF produces the best results in terms of accuracy for all cases including loop intersections and multiple crossing instances. This algorithm also focussed on the processing of various asymmetries in the Gujrati characters as RFMLP. FMRF achieves high successful recognition rate of about 99.8% for all the Gujrati characters.

Finally we conclude the chapter with a note that we are exploring further results on the Gujrati Character Recognition dataset using other soft computing techniques such as fuzzy and rough version of hierarchical bidirectional recurrent neural networks.

Table 9.6 The experimental results for a subset of the Gujrati characters using FMRFs

Characters	Successful Recognition (%)	Unsuccessful Recognition (%)	No Recognition (%)
ખ	99.8	0.2	0
આ	99.8	0.2	0
ઇ	99.8	0.2	0
ઈ	99.8	0.2	0
ઉ	99.8	0.2	0
ઊ	99.8	0.2	0
એ	99.8	0.2	0
ઐ	99.8	0.2	0
ઓ	99.8	0.2	0
ઔ	99.8	0.2	0
ઋ	99.8	0.2	0
ઓ	99.8	0.2	0
ક	99.8	0.2	0
ગ	99.8	0.2	0
ઘ	99.8	0.2	0
ડ	99.8	0.2	0
ચ	99.8	0.2	0
છ	99.8	0.2	0
જ	99.8	0.2	0
ઝ	99.8	0.2	0
ગ	99.8	0.2	0
ય	99.8	0.2	0
શ	99.8	0.2	0
ટ	99.8	0.2	0
ઠ	99.8	0.2	0
ડ	99.8	0.2	0
ઢ	99.8	0.2	0
ણ	99.8	0.2	0
ર	99.8	0.2	0
ષ	99.8	0.2	0
ત	99.8	0.2	0
થ	99.8	0.2	0
દ	99.8	0.2	0
ધ	99.8	0.2	0
ન	99.8	0.2	0
લ	99.8	0.2	0
સ	99.8	0.2	0
પ	99.8	0.2	0
ફ	99.8	0.2	0
બ	99.8	0.2	0
ભ	99.8	0.2	0
મ	99.8	0.2	0
વ	99.8	0.2	0

References

1. Chaudhuri, A., Some Experiments on Optical Character Recognition Systems for different Languages using Soft Computing Techniques, Technical Report, Birla Institute of Technology Mesra, Patna Campus, India, 2010.

2. https://en.wikipedia.org/wiki/Gujarati_language.
3. Bunke, H., Wang, P. S. P. (Editors), Handbook of Character Recognition and Document Image Analysis, World Scientific, 1997.
4. http://www.indsenz.com/int/index.php?content=software_ind_ocr_gujarati.
5. Bansal, V., Sinha, R. M. K., Integrating Knowledge Sources in Devanagari Text Recognition System, IEEE Transactions on Systems, Man, and Cybernetics - Part A: Systems and Humans, 30(4), pp 500–505, 2000.
6. Bajaj, R., Dey, L., Chaudhury, S., Devnagari Numeral Recognition by combining Decision of Multiple Connectionist Classifiers, Sadhana, 27(1), pp 59–72, 2002.
7. http://varamozhi.sourceforge.net/iscii91.pdf.
8. http://homepages.cwi.nl/~dik/english/codes/indic.html.
9. http://acharya.iitm.ac.in/multi_sys/exist_codes.html.
10. Cheriet, M., Kharma, N., Liu, C. L., Suen, C. Y., Character Recognition Systems: A Guide for Students and Practitioners, John Wiley and Sons, 2007.
11. Sharma, N., Pal, U., Kimura, F., Pal, S., Recognition of Offline Handwritten Devnagari Characters using Quadratic Classifier, Indian Conference on Computer Vision, Graphics and Image Processing, pp 805–816, 2006.
12. Gujrati Character Recognition Dataset: Antani, S., Agnihotri, L., Gujrati Character Recognition, Proceedings of the Fifth International Conference on Document Analysis and Recognition, 418, 1999. http://dl.acm.org/citation.cfm?id=840401.
13. Chaudhuri, A., De, K., Job Scheduling using Rough Fuzzy Multi-Layer Perception Networks, Journal of Artificial Intelligence: Theory and Applications, 1(1), pp 4–19, 2010.
14. Chaudhuri, A., De, K., Chatterjee, D., Discovering Stock Price Prediction Rules of Bombay Stock Exchange using Rough Fuzzy Multi-Layer Perception Networks, Book Chapter: Forecasting Financial Markets in India, Rudra P. Pradhan, Indian Institute of Technology Kharagpur, (Editor), Allied Publishers, India, pp 69–96, 2009.
15. Pal, S. K., Mitra, S., Mitra, P., Rough-Fuzzy Multilayer Perception: Modular Evolution, Rule Generation and Evaluation, IEEE Transactions on Knowledge and Data Engineering, 15(1), pp 14–25, 2003.
16. Pal, U., Chaudhuri, B. B., Printed Devnagari script OCR system, Vivek, 10(1), pp 12–24, 1997.
17. Chaudhuri, A., Modified Fuzzy Support Vector Machine for Credit Approval Classification, AI Communications, 27(2), pp 189–211, 2014.
18. Chaudhuri, A., De, Fuzzy Support Vector Machine for Bankruptcy Prediction, Applied Soft Computing, 11(2), pp 2472–2486, 2011.
19. Chaudhuri, A., Fuzzy Rough Support Vector Machine for Data Classification, International Journal of Fuzzy System Applications, 5(2), pp 26–53, 2016.
20. Chaudhuri, A., Applications of Support Vector Machines in Engineering and Science, Technical Report, Birla Institute of Technology Mesra, Patna Campus, India, 2011.
21. Zeng, J., Liu, Z. Q., Type-2 Fuzzy Markov Random Fields and their Application to Handwritten Chinese Character Recognition, IEEE Transactions on Fuzzy Systems, 16(3), pp 747–760, 2008.
22. Taghva, K., Borsack, J., Condit, A., Effects of OCR Errors on Ranking and Feedback using the Vector Space Model, Information Processing and Management, 32(3), pp 317–327, 1996.
23. Taghva, K., Borsack, J., Condit, A., Evaluation of Model Based Retrieval Effectiveness with OCR Text, ACM Transactions on Information Systems, 14(1), pp 64–93, 1996.
24. Taghva, K., Borsack, J., Condit, A., Erva, S., The Effects of Noisy Data on Text Retrieval, Journal of American Society for Information Science, 45 (1), pp 50–58, 1994.
25. Jain, A. K., Fundamentals of Digital Image Processing, Prentice Hall, India, 2006.
26. Russ, J. C., The Image Processing Handbook, CRC Press, 6th Edition, 2011.
27. Young, T. Y., Fu, K. S., Handbook of Pattern Recognition and Image Processing, Academic Press, 1986.

28. De, R. K., Pal, N. R., Pal, S. K., Feature Analysis: Neural Network and Fuzzy Set Theoretic Approaches, Pattern Recognition, 30(10), pp 1579–1590, 1997.
29. Gonzalez, R. C., Woods, R. E., Digital Image Processing, 3rd Edition, Pearson, 2013.
30. De, R. K., Basak, J., Pal, S. K., Neuro-Fuzzy Feature Evaluation with Theoretical Analysis, Neural Networks, 12(10), pp 1429–1455, 1999.
31. Zadeh, L. A., Fuzzy Sets, Information and Control, 8(3), pp 338–353, 1965.
32. Zimmermann, H. J., Fuzzy Set Theory and its Applications, 4th Edition, Kluwer Academic Publishers, Boston, 2001.

Chapter 10
Summary and Future Research

10.1 Summary

This research monograph is the outcome of the technical report *some experiments on optical character recognition systems for different languages using soft computing techniques* [5] from the research work done at Birla Institute of Technology Mesra, Patna Campus, India. All the theories and results are adopted from [5] and compiled in this book. The experiments are performed on several real life character datasets using the MATLAB optimization toolbox [12, 20]. The book is primarily directed towards the students of postgraduate as well as research level courses in pattern recognition, optical character recognition and soft computing in universities and institutions across the globe. A basic knowledge of algebra and calculus is prerequisite in understanding the different concepts illustrated in the book. The book is immensely beneficial to researchers in optical character recognition [5]. The book is also useful for professionals in software industries and several research laboratories interested in studying different aspects of optical character recognition.

The machine simulation of human functions has been a very active and challenging research field since the advent of the digital computers. There are certain areas which require appreciable amount of intelligence, such as number crunching or chess playing where stupendous computational improvements by the digital computers have been achieved. However, humans still outperform even the most powerful computers in relatively routine functions such as vision and object recognition. The machine simulation of human reading involving text and characters is one of these areas which has been the subject of intensive research for the last few decades. Yet it is still far away from the final frontier. In this direction, optical character recognition (OCR) covers all types of machine recognition tasks of different character datasets in various application domains involving passport documents, invoices, bank statements, computerized receipts, business cards, mail, printouts etc. The availability of huge datasets in several languages has created an opportunity to analyse OCR systems analytically and propose different computing models. The prima face in these systems is the recognition accuracy.

© Springer International Publishing AG 2017
A. Chaudhuri et al., *Optical Character Recognition Systems for Different Languages with Soft Computing*, Studies in Fuzziness and Soft Computing 352,
DOI 10.1007/978-3-319-50252-6_10

The inherent degree of vagueness and impreciseness present in reallife character data is resolved by treating the recognition systems through fuzzy and rough sets encompassing indeterminate uncertainty. These uncertainty techniques form the basic mathematical foundation for different soft computing tools. A comprehensive assessment of the proposed methods are performed in terms of the English, French, German, Latin, Hindi and Gujrati languages. The simulation studies have revealed that the soft computing based modeling of OCR systems performs consistently better than traditional models [5]. Several case studies are also presented to show benefits of soft computing models for OCR systems.

In this monograph we have presented some robust soft computing models [1–4, 6, 7, 13, 18] for OCR systems. The experimental framework revolves around six different widely spoken languages across the globe viz English, French, German, Latin, Hindi and Gujrati languages [5] giving due consideration to uncertainty in the character datasets. The inherent uncertainty in the character data are indeterminate in nature which encompass several types of membership degrees [17]. A comprehensive evaluation of the proposed methods are performed with respect to various criteria. In Chaps. 2 and 3 we present the basic concepts of OCR systems [8] and soft computing methods [11, 16] respectively for the readers. Chapters 2 and 3 set the tone for the rest of the Chapters in the book.

In Chap. 2 a brief background and history of OCR systems are presented. The different techniques of OCR systems are described elaborately. Some of the commonly used applications of OCR systems are discussed. The current status and future of the OCR systems are also highlighted.

Chapter 3 starts with a discussion on the different soft computing constituents like fuzzy sets [17], artificial neural networks [10], genetic algorithms [9] and rough sets [14]. The Hough transform and genetic algorithms for fuzzy feature extraction and feature selection [5] respectively are presented next. Finally the soft computing tools like rough fuzzy multilayer perceptron (RFMLP) [6, 7, 13] fuzzy support vector machine (FSVM) [3, 4] and fuzzy rough support vector machine (FRSVM) [2], hierarchical fuzzy bidirectional recurrent neural networks (HFBRNN) [1] and fuzzy markov random fields (FMRF) [18] are described for the readers.

In Chap. 4 the OCR systems for English language are investigated. The different pre-processing activities such as binarization, noise removal, skew detection and correction, character segmentation and thinning are performed on the considered datasets [21]. The feature extraction is performed through discrete cosine transformation. The feature based classification is performed through fuzzy multilayer perceptron (FMLP) [13], RFMLP [6, 7] and two SVM based methods such as FSVM [3, 4] and FRSVM [2]. The high recognition accuracy in the experimental results demonstrate the superiority of soft computing techniques.

In Chap. 5 the OCR systems for French language are investigated. The different pre-processing activities such as text region extraction, skew detection and correction, binarization, noise removal, character segmentation and thinning are performed on the considered datasets [22]. The feature extraction is performed through fuzzy Hough transform. The feature based classification is performed through RFMLP [6, 7, 13] two SVM based methods such as FSVM [3, 4] and FRSVM [2] and HFBRNN [1]. The high recognition accuracy in the experimental results demonstrate the superiority of soft computing techniques.

In Chap. 6 the OCR systems for German language are investigated. The different pre-processing activities such as text region extraction, skew detection and correction, binarization, noise removal, character segmentation and thinning are performed on the considered datasets [23]. The feature extraction is performed through fuzzy genetic algorithms. The feature based classification is performed through RFMLP [6, 7, 13] two SVM based methods such as FSVM [3, 4] and FRSVM [2] and HFBRNN [1]. The high recognition accuracy in the experimental results demonstrate the superiority of soft computing techniques.

In Chap. 7 the OCR systems for Latin language are investigated. The different pre-processing activities such as text region extraction, skew detection and correction, binarization, noise removal, character segmentation and thinning are performed on the considered datasets [23]. The feature extraction is performed through fuzzy genetic algorithms. The feature based classification is performed through RFMLP [6, 7, 13] two SVM based methods such as FSVM [3, 4] and FRSVM [2] and HFBRNN [1]. The high recognition accuracy in the experimental results demonstrate the superiority of soft computing techniques.

In Chap. 8 the OCR systems for Hindi language are investigated. The different pre-processing activities such as binarization, noise removal, skew detection and correction, character segmentation and thinning are performed on the considered datasets [24]. The feature extraction is performed through discrete Hough transformation. The feature based classification is performed through RFMLP [6, 7, 13] two SVM based methods such as FSVM [3, 4] and FRSVM [2] and FMRF [18]. The high recognition accuracy in the experimental results demonstrate the superiority of soft computing techniques.

In Chap. 9 the OCR systems for Gujrati language are investigated. The different pre-processing activities such as binarization, noise removal, skew detection and correction, character segmentation and thinning are performed on the considered datasets [25]. The feature extraction is performed through discrete Hough transformation. The feature based classification is performed through RFMLP [6, 7, 13] two SVM based methods such as FSVM [3, 4] and FRSVM [2] and FMRF [18]. The high recognition accuracy in the experimental results demonstrate the superiority of soft computing techniques.

10.2 Future Research

During the course of this research work certain investigations are made. There are other aspects which could not be addressed in the monograph. These are formulated as the topics for future research work:

(i)   An appreciable amount of work can be done on feature selection and feature extraction [15]. Both feature selection and feature extraction forms an integral part towards the success of any OCR system. The choice of appropriate feature selection and feature extraction methods improves the recognition accuracy of potential OCR system.

(ii) The RFMLP can be further fine-tuned to improve the robustness and accuracy [5] of the results. The fuzzy membership function in RFMLP can be replaced by rough fuzzy and fuzzy rough membership functions which gives better results. Also the MLP can be further restructured by replacing the number hidden layers and the nodes per layer for superior classification results.

(iii) The FSVM and FRSVM can be further improved with better fuzzy and fuzzy rough membership functions [19] respectively. Better results can also achieved by fine-tuning the SVM parameters which take care of overfitting and using different kernels.

(iv) The HFBRNN can be further improved by fine-tuning the parameters [1] which will enhance the results' accuracy. Better rough fuzzy and fuzzy rough membership functions will improve the overall quality of the results obtained through HFBRNN.

(v)   The experiments can also be performed with other soft computing based OCR systems [5]. These OCR systems can be developed by using the different soft computing tools in the optimum possible combinations.

(vi) Several simulation results based on other real life character sets for the six different languages considered [5] can be added in the Chaps. 4–9 respectively. These character sets can be adopted from different research groups working on OCR. We are in the process of accumulating such datasets where experiments could be performed.

(vii) We are also working towards the development of soft computing based OCR systems for some other languages such as Spanish, Portuguese, Italian, Russian, Punjabi, Marathi and Nepali.

References

1. Chaudhuri, A., Ghosh, S. K., Sentiment Analysis of Customer Reviews Using Robust Hierarchical Bidirectional Recurrent Neural Network, Book Chapter: Artificial Intelligence Perspectives in Intelligent Systems, Radek Silhavy, Roman Senkerik, Zuzana Kominkova Oplatkova, Petr Silhavy, Zdenka Prokopova, (Editors), Advances in Intelligent Systems and Computing, Springer International Publishing, Switzerland, Volume 464, pp 249–261, 2016.
2. Chaudhuri, A., Fuzzy Rough Support Vector Machine for Data Classification, International Journal of Fuzzy System Applications, 5(2), pp 26–53, 2016.
3. Chaudhuri, A., Modified Fuzzy Support Vector Machine for Credit Approval Classification, AI Communications, 27(2), pp 189–211, 2014.
4. Chaudhuri, A., De, Fuzzy Support Vector Machine for Bankruptcy Prediction, Applied Soft Computing, 11(2), pp 2472–2486, 2011.

5. Chaudhuri, A., Some Experiments on Optical Character Recognition Systems for different Languages using Soft Computing Techniques, Technical Report, Birla Institute of Technology Mesra, Patna Campus, India, 2010.
6. Chaudhuri, A., De, K., Job Scheduling using Rough Fuzzy Multi-Layer Perception Networks, Journal of Artificial Intelligence: Theory and Applications, 1(1), pp 4–19, 2010.
7. Chaudhuri, A., De, K., Chatterjee, D., Discovering Stock Price Prediction Rules of Bombay Stock Exchange using Rough Fuzzy Multi-Layer Perception Networks, Book Chapter: Forecasting Financial Markets in India, Rudra P. Pradhan, Indian Institute of Technology Kharagpur, (Editor), Allied Publishers, India, pp 69–96, 2009.
8. Cheriet, M., Kharma, N., Liu, C. L., Suen, C. Y., Character Recognition Systems: A Guide for Students and Practitioners, John Wiley and Sons, 2007.
9. Goldberg, D. E., Genetic Algorithms in Search, Optimization and Machine Learning, 4th Edition, Pearson Education, New Delhi, 2009.
10. Haykin, S., Neural Networks and Learning Machines, 3rd Edition, Prentice Hall, 2008.
11. Jang, J. S. R., Sun, C. T., Mizutani, E., Neuro-Fuzzy and Soft Computing: A Computational Approach to Learning and Machine Intelligence, Prentice Hall, 1997.
12. Padhy, N. P., Simon, S. P., Soft Computing: With MATLAB Programming, Oxford University Press, 2015.
13. Pal, S. K, Mitra, S., Mitra, P., Rough Fuzzy MLP: Modular Evolution, Rule Generation and Evaluation, IEEE Transactions on Knowledge and Data Engineering, 15 (1), pp 14–25, 2003.
14. Pawlak, Z., Rough Sets, International Journal of Computer and Information Sciences, 11, pp 341–356, 1982.
15. Webb, A. R., Copsey, K. D., Statistical Pattern Recognition, 3rd Edition, Wiley 2011.
16. Zadeh, L. A., Fuzzy Logic, Neural Networks and Soft Computing, Communications of the ACM, 37(3), pp 77–84, 1994.
17. Zadeh, L. A., Fuzzy Sets, Information and Control, 8(3), pp 338–353, 1965.
18. Zeng, J., Liu, Z. Q., Type-2 Fuzzy Markov Random Fields and their Application to Handwritten Chinese Character Recognition, IEEE Transactions on Fuzzy Systems, 16(3), pp 747–760, 2008.
19. Zimmermann, H. J., Fuzzy Set Theory and its Applications, 4th Edition, Kluwer Academic Publishers, Boston, 2001.
20. http://in.mathworks.com/help/vision/optical-character-recognition-ocr.html.
21. http://www.iam.unibe.ch/fki/databases/iam-handwriting-database.
22. http://www.infres.enst.fr/~elc/GRCE/news/IRONOFF.doc.
23. http://www.inftyproject.org/en/database.html.
24. http://lipitk.sourceforge.net/datasets/dvngchardata.htm.
25. Gujrati Character Recognition Dataset: Antani, S., Agnihotri, L., Gujrati Character Recognition, Proceedings of the Fifth International Conference on Document Analysis and Recognition, 418, 1999. http://dl.acm.org/citation.cfm?id=840401.

Index

© Springer International Publishing AG 2017
A. Chaudhuri et al., *Optical Character Recognition Systems for Different Languages with Soft Computing*, Studies in Fuzziness and Soft Computing 352, DOI 10.1007/978-3-319-50252-6

Printed in the United States
By Bookmasters